"十四五"高等职业教育人工智能技术应用系列教材

自然语言
处理技术与应用

武桂梅　林野川　徐　明◎主　编
王丽媛　张国峰　赵　天　么冰玉　张　淼　叶昭晖◎副主编

中国铁道出版社有限公司
CHINA RAILWAY PUBLISHING HOUSE CO., LTD.

内 容 简 介

本书是介绍自然语言处理基础理论知识和典型应用案例的实战类书籍。本书在整体知识结构上,由浅入深地阐述了自然语言处理的完整知识体系;在案例应用上,采用目前主流的编程方式及开发工具,详细介绍了自然语言相关基础理论、分词和字典的应用、典型数据预处理方式、经典自然语言处理模型及算法流程、感知机模型、条件随机场模型、命名实体方法、信息抽取方法、文本聚类方法、文本分类方法、依存语法(句法)分析方法、自然语言处理中深度学习的应用等内容。

本书适合高等职业院校、应用型本科院校作为自然语言处理课程的教学与实训教材,也可供人工智能从业者作为理论与实践结合的参考书。

图书在版编目(CIP)数据

自然语言处理技术与应用 / 武桂梅,林野川,徐明主编.—北京:中国铁道出版社有限公司,2023.4
"十四五"高等职业教育人工智能技术应用系列教材
ISBN 978-7-113-30102-6

Ⅰ. ①自… Ⅱ. ①武…②林…③徐… Ⅲ. ①自然语言处理 - 高等职业教育 - 教材 Ⅳ. ① TP391

中国国家版本馆 CIP 数据核字(2023)第 054659 号

书 名:自然语言处理技术与应用	
作 者:武桂梅 林野川 徐 明	
策 划:祁 云	编辑部电话:(010)63549458
责任编辑:祁 云 徐盼欣	
封面设计:尚明龙	
责任校对:刘 畅	
责任印制:樊启鹏	

出版发行:中国铁道出版社有限公司(100054,北京市西城区右安门西街 8 号)
网　　址:http://www.tdpress.com/51eds/
印　　刷:河北京平诚乾印刷有限公司
版　　次:2023 年 4 月第 1 版 2023 年 4 月第 1 次印刷
开　　本:850 mm×1 168 mm 1/16 印张:12.75 字数:313 千
书　　号:ISBN 978-7-113-30102-6
定　　价:39.80 元

版权所有　侵权必究

凡购买铁道版图书,如有印制质量问题,请与本社教材图书营销部联系调换。电话:(010)63550836
打击盗版举报电话:(010)63549461

前言

党的二十大报告提出:"推动战略性新兴产业融合集群发展,构建新一代信息技术、人工智能、生物技术、新能源、新材料、高端装备、绿色环保等一批新的增长引擎。"从党的二十大报告可以看出,人工智能已经处于国家战略性地位,这令人工智能领域的工作者倍感振奋。在人工智能领域,自然语言理解处在认知智能核心的地位,它的进步会引导知识图谱的进步,会引导用户理解能力的增强,也会进一步推动整个推理能力的发展。自然语言处理的技术会推动人工智能整体的进展,从而使得人工智能技术可以落地实用化。

随着移动互联网的飞速发展,特别是物联网的发展,人与设备的语言交互场景越来越多,并且逐渐成为核心。这种语言的交互不仅包括语音类的,也包括纯文字的。自然语言处理(Natural Language Processing)是指以计算机和编程语言为工具,对人类特有的书面和口头形式的各种类型的自然语言信息进行加工和处理的技术。当然,随着技术的不断发展,其处理领域也出现了跨形态的组合。比如,通过与图像识别技术的结合,可以实现看图说话、在线问答等应用。因此,自然语言处理是一门交叉性的科学,也常被称为计算语言处理(Computing Language)。没有语言,人类的思维就无从谈起,所以自然语言处理体现了智能化的高级别任务和境界。自然语言处理从涉及的内容上看,既有语法分析,也有语义分析。从应用的角度来看,自然语言处理的应用前景是十分广阔的。特别是在信息化时代,自然语言处理的应用包罗万象,如机器翻译、印刷体文字识别、语音识别、信息检索、信息提取和过滤、文本分类、文本聚类、舆情分析和观点挖掘等,涉及的领域包括数据挖掘、机器学习、知识获取、知识进程、语言计算相关的人工智能研究和语言学研究等。

作为一门交叉程度很高的学科,自然语言处理的发展可谓突飞猛进,无论是对自然语言本质的探究,还是落实到实际应用中,在未来必然会有令人期待的惊喜。

本书主要特色包括:本书采用基于具体案例的实战学习方法,在各种自然语言处理任务中引入经典模型作为解决方案,以实战应用为主、理论和公式推导为辅的形式对自然语言处理中的关键技术进行介绍。同时,为了让学生紧跟学术前沿,本书不仅介绍了自然语言处理模型的基本构建模块,还引入了部分学术界的前沿方法。在整体知识结构上,本书由浅入深地讲解了自然语言处理的知识体系，适合没有接触过自然语言处理领域的学生全面了解相关基础知识；在文字讲述和内容展示上，本书由点及面、图文并茂、深入浅出地阐述自然语言处理领域的基本知识，力求帮助学生迅速掌握基础概念。

本书共分12个单元。单元1是绪论部分，重点介绍自然语言处理领域中的基本概念和术语；单元2是分词和字典，从分词、字典树、切分算法、评测指标四个方面对分词和字典的相关理论进行介绍，本单元内容也是后续自然语言处理任务的基础；单元3是数据预处理，从数据清洗、分词处理、特征工程三个方面对数据预处理知识进行了介绍，并重点介绍了如何在文本上执行预处理任务；单元4是语言模型和算法流程，对隐马尔可夫模型的基础知识及应用方法进行了介绍；单元5是感知机，从分类问题出发，引出感知模型，并对感知机的应用方式进行介绍；单元6是条件随机场，围绕条件随机场的基本概念及应用方法进行介绍；单元7～单元11涵盖了自然语言处理的典型子任务，分别介绍了命名实体识别、信息抽取、文本聚类、文本分类、依存语法分析五个自然语言处理领域的典型子任务，这部分内容是自然语言处理高级任务的基础与核心；单元12以深度学习和自然语言处理为主题，围绕如何在自然语言处理中应用深度学习方法展开，并通过机器翻译这一较为复杂的任务介绍学术界的前沿方法。

本书的读者对象包括：第一，打算学习并入门自然语言处理技术的高等职业院校、应用型本科院校在校生；第二，在金融、医疗、新媒体等行业工作且希望应用人工智能解决本行业问题的工程技术人员；第三，已经对人工智能有一定的了解，想要更深入地学习自然语言处理技术的相关人员；第四，信息和计算机科学爱好者。

本书由武桂梅、林野川、徐明任主编，由王丽媛、张国峰、赵天、么冰玉、张淼、叶昭晖任副主编，和中育数据研发团队共同编写完成。

由于编者水平有限，加之时间仓促，书中难免存在疏漏和不足之处，恳请广大读者批评指正。

编　者

2023年1月

目 录

单元1 绪论 1
    1.1 自然语言和编程语言 2
        1.1.1 自然语言简介 2
        1.1.2 编程语言简介 2
        1.1.3 自然语言和编程语言的相通性 3
    1.2 自然语言处理和信息抽取 3
    1.3 自然语言处理和机器学习 4
    1.4 语料库 5
        1.4.1 通用单语语料库 5
        1.4.2 汉英双语平行语料库 6
        1.4.3 任务专门语料库 6
    实战应用一:信贷违约预测 6
    实战应用二:电子病例解析 14
    单元小结 23
    习题 23

单元2 分词和字典 24
    2.1 分词 25
        2.1.1 中文分词的原理和难点 25
        2.1.2 基于字典的中文分词方法 26
        2.1.3 主流的中文分词工具 26
    2.2 字典树 27
        2.2.1 字典树概述 27
        2.2.2 字典树的作用 28
        2.2.3 字典树的简易实现 28
    2.3 切分算法 29
        2.3.1 切分算法概述 29
        2.3.2 完全切分算法 29
        2.3.3 最长匹配算法 30

2.4	评测指标	31
	2.4.1 机器学习中的准确率	31
	2.4.2 机器学习中的精确率、召回率与F-score	31
	2.4.3 NLP中的精确率、召回率和F-score	32

实战应用：使用HanLP词典实现中文分词 33
单元小结 34
习题 34

单元3 数据预处理 35

- 3.1 数据清洗 35
- 3.2 分词处理 36
- 3.3 特征工程 36
  - 3.3.1 处理流程 36
  - 3.3.2 常用的中文文本处理函数 36

实战应用一：英文新闻资讯数据清洗 41
实战应用二：中文新闻资讯数据清洗 43
单元小结 47
习题 48

单元4 语言模型和算法流程 49

- 4.1 隐马尔可夫链和二元语法 49
  - 4.1.1 隐马尔可夫链背景引入 50
  - 4.1.2 隐马尔可夫链定义 50
  - 4.1.3 二元语法 50
- 4.2 中文分词语料库 50
- 4.3 隐马尔可夫模型和序列标注 51
  - 4.3.1 序列标注 51
  - 4.3.2 隐马尔可夫模型 51
  - 4.3.3 统计n元语法 51
  - 4.3.4 加载模型和构建词网 52
  - 4.3.5 误差分析 52

实战应用：基于马尔可夫模型的文本生成器 60
单元小结 63
习题 63

单元5 感知机 64
5.1 分类问题 64
5.2 基于分类的感知机分类 65
5.3 结构化预测问题 69
5.4 基于结构化感知机的中文分词 69
实战应用：使用感知机根据人名实现性别分类 72
单元小结 76
习题 76

单元6 条件随机场 77
6.1 条件随机场描述 77
6.2 CRF++工具 78
6.3 CRF++特征模板 79
6.4 CRF++命令行预测 81
实战应用：基于条件随机场的词性标注 88
单元小结 90
习题 90

单元7 命名实体识别 91
7.1 命名实体和命名实体识别 92
7.1.1 命名实体 92
7.1.2 命名实体识别 92
7.2 基于规则的命名实体识别 92
7.3 基于层叠隐马尔可夫模型的角色标注框架 92
7.4 基于序列标注的命名实体识别 93
实战应用：热点问题命名实体识别 100
单元小结 109
习题 109

单元8 信息抽取 110
8.1 词性标注 111
8.2 关系抽取 111
8.3 信息熵 111
8.3.1 信息熵的概念 111
8.3.2 信息熵的计算 111

8.4 新词提取 112
8.4.1 新词发现 112
8.4.2 短语提取 112
8.4.3 新词提取 112
8.5 关键词提取和词频统计 113
8.5.1 关键词提取 113
8.5.2 词频统计 114
实战应用一：文本关键词提取 116
实战应用二：手机评论标签提取 119
单元小结 123
习题 123

单元9 文本聚类 124
9.1 文本聚类概述 124
9.2 文本聚类特征提取 125
9.2.1 词袋模型 125
9.2.2 词袋模型中的统计指标 126
9.3 k均值算法 127
9.3.1 基本原理 127
9.3.2 k均值算法的简易实现 128
9.4 重复二分聚类算法 129
9.4.1 基本原理 129
9.4.2 算法实现 130
实战应用一：食品安全评论聚类 131
实战应用二：汽车竞品分析 133
单元小结 140
习题 140

单元10 文本分类 141
10.1 文本分类概述 142
10.2 文本分类特征提取 142
10.2.1 文档向量化 142
10.2.2 文档特征筛选 143
10.3 朴素贝叶斯分类算法 143
10.4 支持向量机分类算法 144

10.5 常用文本分类方法 145
10.6 情感分析 153
实战应用：新闻标题分类 155
单元小结 161
习题 161

单元11 依存语法分析 162

11.1 短语结构树 163
11.2 依存语法树 163
11.3 基于转移的依存语法分析 164
    11.3.1 基于转移的思想 164
    11.3.2 arc-eager方法 164
    11.3.3 基于深度学习的方法 165
实战应用：基于依存语法树的意见抽取 175
单元小结 176
习题 177

单元12 深度学习与自然语言处理 178

12.1 传统方法与深度学习方法 179
    12.1.1 传统方法 179
    12.1.2 深度学习方法 179
12.2 word2vec（词向量） 180
    12.2.1 word2vec算法流程 180
    12.2.2 目标函数 180
    12.2.3 预测函数 181
    12.2.4 模型优化 181
    12.2.5 梯度导数 182
12.3 机器翻译和BERT模型 182
实战应用：使用神经网络实现英文-中文翻译 186
单元小结 192
习题 192

参考文献 193

单元1 绪 论

随着大数据和计算机硬件算力水平的不断发展,人工智能的各种高新前沿技术产业迅速成长并在各个领域内普遍应用。特别是在自然语言处理领域,机器学习和深度学习的算法模型一次次刷新分类和翻译任务的评测指标,甚至超越了人类的水平,由此基于深度学习的自然语言处理在全球范围内掀起了高涨的浪潮。

自然语言处理(Natural Language Processing,NLP)是利用计算机来理解人类语言的领域,涉及利用计算机对大量的自然语言数据进行分析,以收集数据的意义和价值,供实际应用时使用。尽管NLP自20世纪50年代就已经存在,但随着机器学习和深度学习的快速发展,该领域的实际应用才有了巨大进展。本书的大部分内容将重点研究自然语言处理的各种实际应用(如机器范围)以及自然语言处理(如信息提取),并特别着重于实战应用。本单元将首先介绍自然语言处理领域中的基本概念和术语,并通过两个实战应用使读者对于自然语言处理的处理流程建立一个直观的认识。本单元的知识导图如图1.1所示。

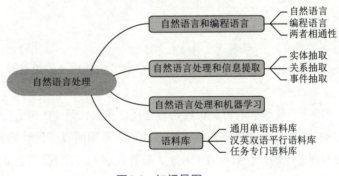

图1.1 知识导图

课程安排

课程任务	课程目标	安排学时
了解自然语言处理	了解自然语言处理定义及基本概念,掌握自然语言和机器语言的共性及联系	1
了解自然语言信息抽取任务	了解信息抽取的基本概念及相关子任务并能有效区分	1
掌握自然语言处理与机器学习关系	对于机器学习与自然语言处理发展有一定认识,对于两者关系有自己的理解	1
熟悉语料库分类及常用语料库	熟悉常用语料库分类并能根据任务需求选择合适的语料库	1
实战应用	通过该实战应用引导读者了解一些业务背景,解决实际问题,进行自我练习、自我提高	2

1.1 自然语言和编程语言

1.1.1 自然语言简介

自然语言（Natural Language）通常是指一种自然地随文化演化的语言。例如，汉语、英语、日语都是自然语言的例子。以一篇汉语文章为例，文章由若干自然段组成，每个自然段中包含若干句子，句子又由词语组成，而文字又是组成词语的基本单位。并且，一篇文章中的语义是由所有的句子语义结合而成的，词语的词义也构建成了整个句子的含义，文字的字面意思相结合就形成了词语。自然语言的基本结构如图 1.2 所示。

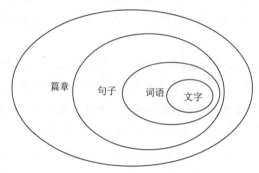

图1.2 自然语言的基本结构

1.1.2 编程语言简介

编程语言（Programming Language）是计算机科学家发明的用来操纵计算机的语言。根据其发展历史可分为：

（1）机器语言，即第一代编程语言。机器语言是用二进制代码（0 和 1）表示的计算机能直接识别和执行的一种机器指令的集合，能够直接被机器执行。早期的程序设计均使用机器语言，程序员将用 0，1 数字编成的程序代码打在纸带或卡片上，打孔代表 1，不打孔代表 0，再将程序通过纸带机或卡片机输入计算机，进行运算。

（2）汇编语言，即第二代编程语言。其使用一些容易理解和记忆的缩写单词来代替一些特定的指令，例如，用 "ADD" 代表加法操作指令，"SUB" 代表减法操作指令，"INC" 代表增加 1，"DEC" 代表减去 1，"MOV" 代表变量传递，等等。助记符的使用大大提高了语言的可读性，使程序的修改和维护都变得更加方便。由于仍然保持了机器语言优秀的执行效率，汇编语言到现在依然是常用的编程语言之一。

（3）高级语言，即第三代编程语言。其可以对多条指令进行整合，将其变为单条指令完成输送，其在操作细节指令以及中间过程等方面都得到了适当的简化，所以，整个程序更为简便，具有较强的操作性，而这种编码方式的简化，使得计算机编程对于相关工作人员的专业水平要求不断放宽。

习惯上将机器语言和汇编语言称为低级语言，另外还有脚本语言和专用语言等。编程语言基本分类如图 1.3 所示。

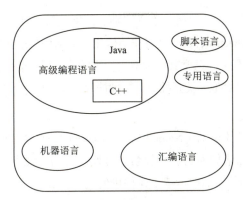

图1.3 编程语言基本分类

1.1.3 自然语言和编程语言的相通性

1. 相似的语法结构和实质作用

自然语言和计算机编程语言之间的句法、语法、词法和语义都有着自己的一整套表达方式。单词和语句简洁是 C 语言的特点。如果将英语和 C 语言进行对比,会发现其中的字词语段在 C 语言当中都是存在的,有固定的篇章和概念,也对对应的语法规则进行了规范。

2. 编程语言随自然语言而诞生

在每一门计算机编程语言诞生之前,这门语言的设计者在设计的过程中会根据自己的母语进行设计。从最初的 Fortran 语言到后来的 Python 语言,对其过程进行分析就可以看到,设计者会将自然语言当中的某些机制并入编程语言中去。现在所有主流编程语言都是用英文拼写的。

3. 发展多样化

当今世界大约存在 6 000 余种自然语言,其中有书面文字的语言有 2 000 多种。而编程语言的数量也非常多,其中大多数都是高级语言,它们在不同的应用领域中被使用。其根据设计思想可被分为命令式语言、函数式语言、面向对象语言等。在工业生产中最常用的高级语言是 C++、Java、Python 等。

1.2 自然语言处理和信息抽取

为了便于理解,将"自然语言处理"这个术语分为两部分解释:

(1)自然语言是一种有机且自然发展而来的书面和口头交流形式。

(2)处理是指用计算机分析和理解输入数据。

如图 1.4 所示,自然语言处理是人类语言的计算机处理,旨在使计算机理解和处理人类的语言,从而在人与计算机之间建立一个简单的沟通关系,使人类能更方便地操纵计算机进行生产工作。

自然语言处理的工作大致可划分为下面几个步骤:首先,接收人类的自然语言;其次,通过信息提取,将自然语言转换为结构化数据;最后,通过分析数据得出结果并输出。自然语言处理应用广泛,其中

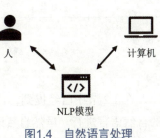

图1.4 自然语言处理

包括人机对话交流、语言翻译、情感分析、语音识别、搜索引擎、社交网站推送等。

信息抽取（INformation Extraction，IE）是将非结构化或半结构化描述的自然语言文本转化成结构化特征的一种基础自然语言处理任务。通常其在抽取出特定的信息后，会保存到结构化的数据库当中，以便用户查询和使用。它包括实体抽取、关系抽取、事件抽取三类子任务。

1. 实体抽取

实体抽取任务又名命名实体识别（Named Entity Recognition，NER）。其是信息抽取的基础性工作，其任务是从文本中识别出诸如人名、组织名、日期、时间、地点、特定的数字形式等内容，并为之添加相应的标注信息，为信息抽取后续工作提供便利。

2. 关系抽取

关系抽取的作用是获取文本中实体之间存在的语法或语义上的联系。关系抽取是信息抽取中的关键环节。例如，在图1.5的句子中，"Jackie R. Brown"、"Washington"、"United States of America"三个实体两两之间具有"是……的首都"和"出生地"的关系。关系抽取的目标是让计算机自动识别出这些关系。近年来，该领域的研究已从最初的单纯语言学模型的应用发展到使用NLP技术的应用和复杂机器学习方法的应用，而关系抽取的性能也随之有了大幅提升。

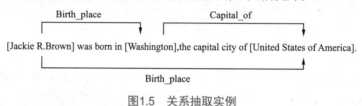

图1.5　关系抽取实例

3. 事件抽取

事件抽取（Event Extraction）主要研究如何从含有事件信息的非结构化文本中抽取出用户感兴趣的事件信息，并将用自然语言表达的事件以结构化的形式呈现出来。这里的"事件"，指若干与特定矛盾相关的事物在某一时空内的运动。简单来说，符合"谁在什么时间在哪做了什么事"等模式的句子，描述的就是一个事件。比如，"我昨天晚上7点去体育馆跑步3 km"这句话，就是在描述一个事件，其结构化描述见表1.1。

表1.1　一次运动的结构化描述信息

序　号	字　段	例　子	序　号	字　段	例　子
1	运动时间	昨晚7点	4	运动地点	体育馆
2	运动者	我	5	运动强度	3 km
3	运动名称	跑步	—	—	—

1.3　自然语言处理和机器学习

在1.2节中已经提到，自然语言处理的任务是建立人与计算机沟通的桥梁，而沟通必须以人类的自然语言进行。虽然自然语言与计算机使用的编程语言具有多个维度的相通性，但想让计算机理解人的意图仍然不是一件简单的事。早在20世纪50年代就有学者提出了自然语言处理这个概念。

当时学术界分为两大阵营：一是基于规则方法的符号派（Symbolic），二是采用概率方法的随机派（Stochastic）。这两大阵营都属于机器学习的范畴，只是前者注重逻辑和推理，后者注重概率与统计。

以 Chomsky 为代表的符号派学者开始了形式语言理论和生成句法的研究，20 世纪 60 年代末又进行了形式逻辑系统的研究。他们的研究成果奠定了后来多种编程语言发展的基础，并对高级语言编译器的发展起到了极大的促进作用。可以说，没有这些研究，就没有今天百花齐放的各种编程语言。

随机派学者采用基于贝叶斯理论的统计学研究方法和神经网络，在这一时期也取得了很大的进步。但受限于当时计算机的性能和存储条件，随机派研究的影响力远不如符号派。在随后的几十年中，使用概率方法处理自然语言一直不温不火，直到 2010 年以来，计算机硬件算力水平迈上了一个新的台阶，为使用基于概率的深度学习技术处理自然语言改善了物质基础。

自然语言处理与机器学习、深度学习的关系如图 1.6 所示。深度学习是近年来机器学习中兴起的一个分支。

深度学习（Deep Learning）是人工智能的深层次理论。深度学习是通过大量训练数据更新模型参数来达到学习训练样本的内部规律以及样本的表示层次。深度学习的神经网络由输入层、隐藏层和输出层构成。深度学习与普通的浅层次学习最大的不同就在于其神经网络的层数更深，通过对输入特征进行逐层变换并向更深层次传播，就能够将特征从原空间转换到一个新的特征空间，从而使数据分析

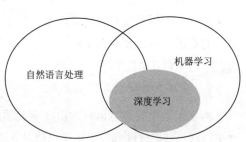

图1.6 自然语言处理与机器学习、深度学习的关系

更加容易。2013 年 word2vec 模型的提出，使得大规模的词嵌入模型训练成为可能。从此深度学习方法在 NLP 领域遍地开花，不断有新的方法被提出，并且在部分子领域中，基于深度学习的神经网络效果已经超过了人类。

1.4 语料库

语料库指经科学取样和加工的大规模电子文本库，其中存放的是在语言的实际使用中真实出现过的语言材料。根据不同的划分标准，语料库可以分为多种类型。例如，按语种划分可以分为单语种语料库和多语种语料库；按记载媒体不同可以分为单媒体语料库和多媒体语料库；按照地域区别可以分为国家语料库和国际语料库等，也有针对具体任务专门构建的语料库。下面列举一些国内具有代表性的语料库。

1.4.1 通用单语语料库

1. 国家语委现代汉语通用平衡语料库

由国家语言文字工作委员会主持，面向语言文字信息处理、语言文字规范和标准的制定、语言文字的学术研究、语文教育以及语言文字的社会应用，总体规模达 1 亿字，语料时间跨度为 1919—2002 年，收录了人文与社会科学、自然科学及综合三个大类约 40 个小类的语料。其中标注语料库为国家语委现代汉语通用平衡语料库全库的子集，该子集按照预先设计的选材原则进行平衡抽样，对

语料进行分词和词类标注，并经过三次人工校对，最后得到约 5 000 万字符的标注语料库。

2. 北京语言大学 BCC 语料库

北京语言大学 BCC 语料库是以汉语为主，兼有英语、西班牙语、法语、德语、土耳其语等语言的语料库，其中汉语语料规模约 150 亿字，涵盖了报刊、文学、微博、科技、综合和古汉语等多领域语料。BCC 语料库包括生语料、分词语料、词性标注语料和句法树，目前已对现代汉语、英语、法语的语料进行词性标注，中、英文句法树则是引自美国宾夕法尼亚大学的中文和英文树库。

1.4.2 汉英双语平行语料库

1. 中国科学院汉英平行语料库

中国科学院汉英平行语料库是在对中英文篇章级对齐的双语文本进行段落对齐、句子对齐加工后建立的一个句子级对齐的双语语言信息和知识库，该语料库借助互联网等媒体搜集中英文篇章级对齐的双语文本，面向多领域、多体裁，采用基于双语词典的句子对齐方法进行文本对齐，并对双语文本句子对齐结果实现自动评价。

2. 清华大学中英平行语料库

清华大学中英平行语料库由清华大学自然语言处理与社会人文计算实验室利用自身研发的互联网平行网页获取软件和双语句子自动对齐软件获取并处理得到的，共包含 285 万中英平行句对。

1.4.3 任务专门语料库

针对分词、词性标注、命名实体识别、词法分析、文本分类、文本聚类等特定任务，都有相对应的公开语料库可供广大研究者使用。比如，中文分词的常用语料库有 MSR、AS、CITYU 等。NLTK、Jieba、SnowNLP 等一些比较有代表性的 NLP 的 Python 库都内含了一些常用的分词、词性标注等常见的语料库。在实际场景中，可以直接通过 API 调用它们，而不需要重新训练分词、词性标注等模型。

实战应用一：信贷违约预测

应用背景：个人信贷业务的需求随着人们生活水平的不断提高和消费观念的转变而不断增长，且在发展过程中暴露了一些问题。因此，应正确认识个贷业务风险。

假设毕业后你进入了一家银行工作，你所在的团队接到一个关于个人信贷预测的任务，要求根据贷款申请人的数据信息预测其违约风险，以此判断是否通过此项贷款，以此来降低银行的金融风控风险。项目经理提供了 120 万条个人信贷数据用于模型的训练，希望团队重视该项目，并尽量完成工作。

1. 数据集

该个人信贷模拟数据集不涉及个人隐私，为计算机模拟数据。其总数据量超过 120 万条。

数据字段见表 1.2。

表1.2 数据字段

字　　段	字段描述
id	为贷款清单分配的唯一信用证标识
loanAmnt	贷款金额
term	贷款期限（年）
interestRate	贷款利率
installment	分期付款金额
grade	贷款等级
subGrade	贷款等级之子级
employmentTitle	就业职称
employmentLength	就业年限（年）
homeOwnership	借款人在登记时提供的房屋所有权状况
annualIncome	年收入
verificationStatus	验证状态
issueDate	贷款发放的月份
purpose	借款人在贷款申请时的贷款用途类别
postCode	借款人在贷款申请中提供的邮政编码的前3位数字
regionCode	地区编码
dti	债务收入比
delinquency_2years	借款人过去2年信用档案中逾期30天以上的违约事件数
ficoRangeLow	借款人在贷款发放时的fico所属的下限范围
ficoRangeHigh	借款人在贷款发放时的fico所属的上限范围
openAcc	借款人信用档案中未结信用额度的数量
pubRec	贬损公共记录的数量
pubRecBankruptcies	公开记录清除的数量
revolBal	信贷周转余额合计
revolUtil	循环额度利用率，或借款人使用的相对于所有可用循环信贷的信贷金额
totalAcc	借款人信用档案中当前的信用额度总数
initialListStatus	贷款的初始列表状态
applicationType	表明贷款是个人申请还是多个共同借款人的联合申请
earliesCreditLine	借款人最早报告的信用额度开立的月份
title	借款人提供的贷款名称
policyCode	公开可用的策略_代码=1，新产品不公开可用的策略_代码=2
n系列匿名特征	匿名特征n0～n14，为一些贷款人行为计数特征的处理

2. 数据读取及分析

```
## 数据读取
import pandas as pd      # 数据分布统计
df = pd.read_csv("/train.csv")
```

```
test = pd.read_csv("/testA.csv")
df.shape

# 查看重复值
df[df.duplicated()==True]# 打印重复值

# 统计关键变量比例
(df['isDefault'].value_counts()/len(df)).round(2)

# 查看统计量
df.describe().T
```

部分字段特征统计量结果如图1.7所示。

loanAmnt	800000	14416.81888	8716.086178	500	8000	12000	20000	40000
term	800000	3.482745	0.855832	3	3	3	3	5
interestRate	800000	13.238391	4.765757	5.31	9.75	12.74	15.99	30.99
installment	800000	437.947723	261.460393	15.69	248.45	375.135	580.71	1715.42
employmentTitle	799999	72005.35171	106585.6402	0	427	7755	117663.5	378351
homeownership	800000	0.614213	0.675749	0	0	1	1	5
annualincome	800000	76133.91049	68947.51367	0	45600	65000	90000	10999200
verification Status	800000	1.009683	0.782716	0	0	1	2	2
isDefault	800000	0.199513	0.399634	0	0	0	0	1
purpose	800000	1.745982	2.367453	0	0	0	4	13
postCode	799999	258.535648	200.037446	0	103	203	395	940
regionCode	800000	16.385758	11.036679	0	8	14	22	50
dti	799761	18.284557	11.150155	-1	11.79	17.61	24.06	999
delinquency_2years	800000	0.318239	0.880325	0	0	0	0	39
ficoRangeLow	800000	696.204081	31.865995	630	670	690	710	845
ficoRangeHigh	800000	700.204226	31.866674	634	674	694	714	850
openAcc	800000	11.59802	5.475286	0	8	11	14	86
pubRec	800000	0.214915	0.606467	0	0	0	0	86
pubRecBankruptcies	799595	0.134163	0.377471	0	0	0	0	12
revolBal	800000	16228.70651	22458.02054	0	5944	11132	19734	2904836
revolUtil	799469	51.790734	24.516126	0	33.4	52.1	70.7	892.3
totalAcc	800000	24.998861	11.999201	2	16	23	32	162
initialList Status	800000	0.416953	0.493055	0	0	0	1	1
applicationType	800000	0.019267	0.137464	0	0	0	0	1

图1.7　部分字段特征统计量结果

通过特征量统计可以确定特征缺失值以及特征的标准差。

```
# 变量种类统计
df.nunique()
df = df.drop(['id','policyCode'],axis=1) # 删除ID列及只有一个值的policyCode列


# 查看特征分布一致性
# 分离数值变量与分类变量
```

```python
Nu_feature = list(df.select_dtypes(exclude=['object']).columns)  # 数值变量
Ca_feature = list(df.select_dtypes(include=['object']).columns)
# 查看数值型训练集与测试集分布
Nu_feature.remove('isDefault')  # 移除目标变量
# 画图
import matplotlib.pyplot as plt
import seaborn as sns
import warnings
warnings.filterwarnings("ignore")
plt.figure(figsize=(30,30))
i=1
for col in Nu_feature:
    ax=plt.subplot(8,5,i)
    ax=sns.distplot(df[col],color='violet')
    ax=sns.distplot(test[col],color='lime')
    ax.set_xlabel(col)
    ax.set_ylabel('Frequency')
    ax=ax.legend(['train','test'])
    i+=1
plt.show()

# 数据相关性查看
plt.figure(figsize=(10,8))
train_corr=df.corr()
sns.heatmap(train_corr,vmax=0.8,linewidths=0.05,cmap="Blues")
```

相关性结果分析如图 1.8 所示。

请读者自行分析哪些变量相关性较高以及目标变量与特征变量的相关性。

3. 数据清洗

```python
# 分类变量处理
from sklearn.preprocessing import LabelEncoder
lb = LabelEncoder()
cols = ['grade','subGrade']
for j in cols:
    df[j] = lb.fit_transform(df[j])
df[cols].head()

# grade 及 subGrade 是有严格的字母顺序的，与测试集相对应，可以直接用编码转换，转换结果如下
  grade  subGrade
0   4      21
1   3      16
2   3      17
3   0      3
4   2      11

# 年限转化为数字，再进行缺失值填充
```

```python
df['employmentLength']=df['employmentLength'].str.replace(' years','').str.replace(' year','').str.replace('+','').replace('< 1',0)

# 随机森林填补年限缺失值,由于分类变量只有年限有缺失,所以这样填充
from sklearn.tree import DecisionTreeClassifier
DTC = DecisionTreeClassifier()
empLenNotNull = df.employmentLength.notnull()
columns = ['loanAmnt','grade','interestRate','annualIncome','homeOwnership','term','regionCode']
# regionCode 变量加入后,准确度从 0.85 提升至 0.97
DTC.fit(df.loc[empLenNotNull,columns], df.employmentLength[empLenNotNull])
print(DTC.score(df.loc[empLenNotNull,columns], df.employmentLength[empLenNotNull]))
# DTC.score: 0.9828872204324179

# 填充
for data in [df]:
    empLen_pred = DTC.predict(data.loc[:,columns])      # 对年限数据进行预测
    empLenIsNull = data.employmentLength.isnull()       # 判断是否为空值,isnull 返回的是布尔值
    data.employmentLength[empLenIsNull] = empLen_pred[empLenIsNull] # 如果是空值,则进行填充

# 转化为整数
df['employmentLength']=df['employmentLength'].astype('int64')

import datetime
df['issueDate']=pd.to_datetime(df['issueDate'])
df['issueDate_year']=df['issueDate'].dt.year.astype('int64')
df['issueDate_month']=df['issueDate'].dt.month.astype('int64')
df['earliesCreditLine']=pd.to_datetime(df['earliesCreditLine'])   # 先在 Excel 上转化为日期
df['earliesCreditLine_year']=df['earliesCreditLine'].dt.year.astype('int64')
df['earliesCreditLine_month']=df['earliesCreditLine'].dt.month.astype('int64')
df=df.drop(['issueDate','earliesCreditLine'],axis=1)
# issueDate 及 earliesCreditLine 两个变量将日期分解,分别提取"年"和"月"并转化为整数便于计算,由于测试集这两个变量的"日"都是 1,对目标变量没有影响,所以训练集不提取,提取完后将这两个原始变量删除

# 数值变量填充
df[Nu_feature] = df[Nu_feature].fillna(df[Nu_feature].median())
# 考虑平均值易受极值影响,数值变量用中位数填充
# 数据保存
df.to_csv("/df2.csv")
```

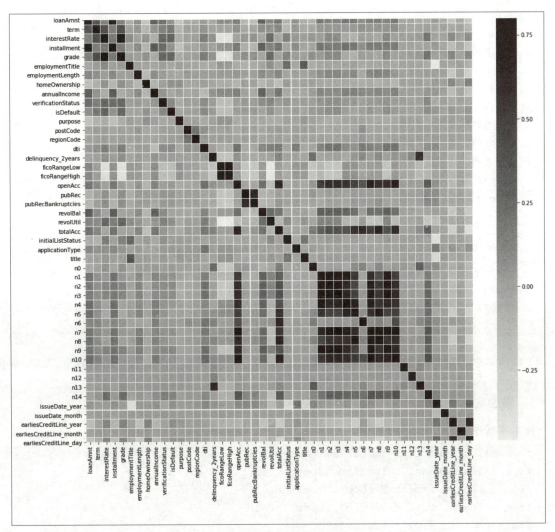

图1.8 相关性分析结果

4. 特征工程

```
# 主成分分析
from sklearn.decomposition import PCA
pca = PCA()
X1=df2.drop(columns='isDefault')
df_pca_train = pca.fit_transform(X1)
pca_var_ration = pca.explained_variance_ratio_
pca_cumsum_var_ration = np.cumsum(pca.explained_variance_ratio_)
print("PCA 累计解释方差 ")
print(pca_cumsum_var_ration)
x=range(len(pca_cumsum_var_ration))
plt.scatter(x,pca_cumsum_var_ration)

# 特征筛选
```

```python
import toad
toad_quality = toad.quality(df2, target='isDefault', iv_only=True)
# 计算各种评估指标，如iv值、gini指数，entropy熵，以及unique values，结果按iv值排序

selected_data, drop_lst = toad.selection.select(df2,target = 'isDefault',
empty = 0.5, iv = 0.02, corr = 0.7,return_drop=True)
# 筛选空值率>0.5、IV<0.02、相关性大于0.7的特征
# (800000, 15) 保留了15个特征
# 以下是删除的特征，通过return_drop=True显示
    {'empty': array([], dtype=float64),
     'iv': array(['employmentLength', 'purpose', 'postCode', 'regionCode',
        'delinquency_2years', 'openAcc', 'pubRec', 'pubRecBankruptcies',
        'revolBal', 'totalAcc', 'initialListStatus', 'applicationType',
        'n0', 'n1', 'n4', 'n5', 'n6', 'n7', 'n8', 'n10', 'n11', 'n12',
        'n13', 'issueDate_month', 'earliesCreditLine_year',
        'earliesCreditLine_month'], dtype=object),
     'corr': array(['n9', 'grade', 'n3', 'installment', 'ficoRangeHigh',
        'interestRate'], dtype=object)}

psi = toad.metrics.PSI(df2,testA)     # psi没有大于0.25的，都比较稳定
psi.sort_values(0,ascending=False)
```

5. 模型构建

在本实践中，选用catboost模型。

```python
from sklearn.metrics import roc_auc_score
from sklearn.model_selection import train_test_split
from catboost import CatBoostClassifier
from sklearn.model_selection import KFold
train=pd.read_csv("/df2.csv")
testA2=pd.read_csv("/testA.csv")
# 选取相关变量做分类变量并转化为字符串格式
col=['grade','subGrade','employmentTitle','homeOwnership','verificationStatus','purpose','issueDate_year','postCode','regionCode','earliesCreditLine_year','issueDate_month','earliesCreditLine_month','initialListStatus','applicationType']
for i in train.columns:
    if i in col:
        train[i] = train[i].astype('str')
for i in testA2.columns:
    if i in col:
        testA2[i] = testA2[i].astype('str')
# 划分特征变量与目标变量
X=train.drop(columns='isDefault')
Y=train['isDefault']
# 划分训练及测试集
x_train,x_test,y_train,y_test=train_test_split(X,Y,test_size=0.2,random_state=123)
```

```
# 模型训练
clf=CatBoostClassifier(
            loss_function="Logloss",
            eval_metric="AUC",
            task_type="CPU",
            learning_rate=0.1,
            iterations=300,
            random_seed=2022,
            od_type="Iter",
            depth=7)
result = []
mean_score = 0
n_folds=3
kf = KFold(n_splits=n_folds ,shuffle=True,random_state=2022)
for train_index, test_index in kf.split(X):
    x_train = X.iloc[train_index]
    y_train = Y.iloc[train_index]
    x_test = X.iloc[test_index]
    y_test = Y.iloc[test_index]
    clf.fit(x_train,y_train,verbose=300,cat_features=col)
    y_pred=clf.predict_proba(x_test)[:,1]
    print('验证集auc:{}'.format(roc_auc_score(y_test, y_pred)))
    mean_score += roc_auc_score(y_test, y_pred) / n_folds
    y_pred_final = clf.predict_proba(testA2)[:,-1]
    result.append(y_pred_final)
```

6. 模型评估

```
# 模型评估
print('mean 验证集 Auc:{}'.format(mean_score))
cat_pre=sum(result)/n_folds
```

模型训练及验证结果如图 1.9 所示。

```
0:    total: 3.13s    remaining: 15m 35s
299:  total: 9m 15s remaining: 0us
验证集auc:0.7388007571702323
0:    total: 2.08s    remaining: 10m 20s
299:  total: 9m 45s remaining: 0us
验证集auc:0.7374681864389327
0:    total: 1.73s    remaining: 8m 38s
299:  total: 9m 22s remaining: 0us
验证集auc:0.7402961974320663
mean 验证集Auc:0.7388550470137438
```

图1.9 模型训练及验证结果

本实战应用是一个典型的机器学习应用处理问题，希望通过本次实战，同学们对如何用编程语言解决实际问题有一个整体认识，对特征工程有一定了解，接下来的单元中会介绍如何解决自然语言处理问题。

实战应用二：电子病例解析

应用背景：

电子病历是现代化医院质量管理和病历管理的必然趋势，是医院信息系统的核心，是计算机技术和网络技术在医疗领域的必然产物。电子病历对于提高医疗质量管理，提高病历的规范性、完整性，整理共享病人信息，加强对医疗质量的监督，减轻医生的工作量，提高医务人员的工作效率等方面都有非常重要的价值，是医疗质量管理工作的重点之一。

医院客户委托项目组希望能根据患者住院期间的临床记录来预测该患者未来 30 天内是否会再次入院，通过预测辅助医生在选择治疗方案和评估手术风险方面做更好的工作。在临床中，治疗手段常见而愈后情况难以控制管理的情况屡见不鲜。比如，关节置换手术作为治疗老年骨性关节炎等疾病的最终方法在临床中取得了极大成功，但是与手术相关的并发症以及由此导致的再次入院情况并不少见。患者的自身因素如心脏病、糖尿病、肥胖等情况也会增加关节置换术后的再次入院风险。在接受关节置换手术者的年龄越来越大、健康状况越来越差的情况下，会出现更多的并发症，并且增加再次入院风险。

通过电子病历的相关记录，观察到对于某些疾病或者手术来说，30 天内再次入院的患者各方面的风险都明显增加。因此，对与前次住院原因相同，且前次出院与下次入院间隔未超过 30 天的再一次住院视为同一次住院的情况进行筛选标注，训练模型来尝试解决这个问题。

1. 数据集

选取于 Medical Information Mart for Intensive Care Ⅱ 数据集，也称 MIMIC-Ⅱ，是在 NIH 资助下，由 MIT、哈佛医学院 BID 医学中心、飞利浦医疗联合开发维护的多参数重症监护数据库。

2. 环境配置

```
pip install -U pytorch-pretrained-bert -i https://pypi.tuna.tsinghua.edu.cn/simple
from IPython.core.interactiveshell import InteractiveShell
InteractiveShell.ast_node_interactivity='all'
```

3. 数据分析

```
import pandas as pd
sample = pd.read_csv('/home/sample.csv')
```

数据格式见表 1.3。

表1.3 数据格式

序 号	TEXT	ID	Label
0	Nursing Progress Note 1900-0700 hours:\n** Ful...	176088	1
1	Nursing Progress Note 1900-0700 hours:\n** Ful...	135568	1
2	NPN:\n\nNeuro: Alert and oriented X2-3, Sleepi...	188180	0
3	RESPIRATORY CARE:\n\n35 yo m adm from osh for ...	110655	0
4	NEURO: A+OX3 pleasant, mae, following commands...	139362	0

4. 模型构建

```python
import csv
import pandas as pd

class InputExample(object):
    def __init__(self, guid, text_a, text_b=None, label=None):
        self.guid = guid
        self.text_a = text_a
        self.text_b = text_b
        self.label = label


class InputFeatures(object):
    def __init__(self, input_ids, input_mask, segment_ids, label_id):
        self.input_ids = input_ids
        self.input_mask = input_mask
        self.segment_ids = segment_ids
        self.label_id = label_id


class DataProcessor(object):
    def get_labels(self):
        raise NotImplementedError()

    def _read_tsv(cls, input_file, quotechar=None):
        with open(input_file, "r") as f:
            reader = csv.reader(f, delimiter="\t", quotechar=quotechar)
            lines = []
            for line in reader:
                lines.append(line)
            return lines

    def _read_csv(cls, input_file):
        file = pd.read_csv(input_file)
        lines = zip(file.ID, file.TEXT, file.Label)
        return lines
```

```python
def create_examples(lines, set_type):
    examples = []
    for (i, line) in enumerate(lines):
        guid = "%s-%s" % (set_type, i)
        text_a = line[1]
        label = str(int(line[2]))
        examples.append(
            InputExample(guid=guid, text_a=text_a, text_b=None, label=label))
    return examples


class ReadmissionProcessor(DataProcessor):
    def get_test_examples(self, data_dir):
        return create_examples(
            self._read_csv(os.path.join(data_dir, "sample.csv")), "test")

    def get_labels(self):
        return ["0", "1"]


# 将语料对按最大长度截取语料
def truncate_seq_pair(tokens_a, tokens_b, max_length):
    while True:
        total_length = len(tokens_a) + len(tokens_b)
        if total_length <= max_length:
            break
        if len(tokens_a) > len(tokens_b):
            tokens_a.pop()
        else:
            tokens_b.pop()


# 将语料对按最大长度截取语料
def truncate_seq_pair(tokens_a, tokens_b, max_length):
    while True:
        total_length = len(tokens_a) + len(tokens_b)
        if total_length <= max_length:
            break
        if len(tokens_a) > len(tokens_b):
            tokens_a.pop()
        else:
            tokens_b.pop()

    tokens = []
```

```python
segment_ids = []
tokens.append("[CLS]")
segment_ids.append(0)
for token in tokens_a:
    tokens.append(token)
    segment_ids.append(0)
tokens.append("[SEP]")
segment_ids.append(0)

if tokens_b:
    for token in tokens_b:
        tokens.append(token)
        segment_ids.append(1)
    tokens.append("[SEP]")
    segment_ids.append(1)

input_ids = tokenizer.convert_tokens_to_ids(tokens)

input_mask = [1] * len(input_ids)

while len(input_ids) < max_seq_length:
    input_ids.append(0)
    input_mask.append(0)
    segment_ids.append(0)

assert len(input_ids) == max_seq_length
assert len(input_mask) == max_seq_length
assert len(segment_ids) == max_seq_length
# print (example.label)
label_id = label_map[example.label]
if ex_index < 5:
    logger.info("*** Example ***")
    logger.info("guid: %s" % (example.guid))
    logger.info("tokens: %s" % " ".join(
            [str(x) for x in tokens]))
    logger.info("input_ids: %s" % " ".join([str(x) for x in input_ids]))
    logger.info("input_mask: %s" % " ".join([str(x) for x in input_mask]))
    logger.info(
            "segment_ids: %s" % " ".join([str(x) for x in segment_ids]))
    logger.info("label: %s (id = %d)" % (example.label, label_id))

features.append(
        InputFeatures(input_ids=input_ids,
```

```python
                                            input_mask=input_mask,
                                            segment_ids=segment_ids,
                                            label_id=label_id))
        return features

# 准确率曲线与绘图
import numpy as np
from sklearn.metrics import roc_curve, auc
import matplotlib.pyplot as plt

def vote_score(df, score, ax):
    df['pred_score'] = score
    df_sort = df.sort_values(by=['ID'])
    # score
     temp = (df_sort.groupby(['ID'])['pred_score'].agg(max) + df_sort.groupby(['ID'])['pred_score'].agg(sum) / 2) / (
                    1 + df_sort.groupby(['ID'])['pred_score'].agg(len) / 2)
    x = df_sort.groupby(['ID'])['Label'].agg(np.min).values
    df_out = pd.DataFrame({'logits': temp.values, 'ID': x})

    fpr, tpr, thresholds = roc_curve(x, temp.values)
    auc_score = auc(fpr, tpr)

    ax.plot([0, 1], [0, 1], 'k--')
    ax.plot(fpr, tpr, label='Val (area = {:.3f})'.format(auc_score))
    ax.set_xlabel('False positive rate')
    ax.set_ylabel('True positive rate')
    ax.set_title('ROC curve')
    ax.legend(loc='best')
    return fpr, tpr, df_out

from sklearn.metrics import precision_recall_curve
from funcsigs import signature

def pr_curve_plot(y, y_score, ax):
    precision, recall, _ = precision_recall_curve(y, y_score)
    area = auc(recall, precision)
    step_kwargs = ({'step': 'post'}
                   if 'step' in signature(plt.fill_between).parameters
                   else {})

    ax.step(recall, precision, color='b', alpha=0.2,
            where='post')
```

```python
        ax.fill_between(recall, precision, alpha=0.2, color='b', **step_kwargs)
        ax.set_xlabel('Recall')
        ax.set_ylabel('Precision')
        ax.set_ylim([0.0, 1.05])
        ax.set_xlim([0.0, 1.0])
        ax.set_title('Precision-Recall curve: AUC={0:0.2f}'.format(
            area))

    def vote_pr_curve(df, score, ax):
        df['pred_score'] = score
        df_sort = df.sort_values(by=['ID'])
        # score
         temp = (df_sort.groupby(['ID'])['pred_score'].agg(max) + df_sort.
groupby(['ID'])['pred_score'].agg(sum) / 2) /
                (1 + df_sort.groupby(['ID'])['pred_score'].agg(len) / 2)
        y = df_sort.groupby(['ID'])['Label'].agg(np.min).values

        precision, recall, thres = precision_recall_curve(y, temp)
         pr_thres = pd.DataFrame(data=list(zip(precision, recall, thres)),
columns=['prec', 'recall', 'thres'])

        pr_curve_plot(y, temp, ax)

        temp = pr_thres[pr_thres.prec > 0.799999].reset_index()

        rp80 = 0
        if temp.size == 0:
            print('Test Sample too small or RP80=0')
        else:
            rp80 = temp.iloc[0].recall
            print(f'Recall at Precision of 80 is {rp80}')

        return rp80
```

5. 模型训练

```python
# 参数配置
config = {
    "local_rank": -1,
    "no_cuda": False,
    "seed": 42,
    "output_dir": './result',
    "task_name": 'readmission',
    "bert_model": '/home/kesci/input/MIMIC_note3519/BERT/early_readmission',
    "fp16": False,
```

```python
        "data_dir": '/home/kesci/input/MIMIC_note3519/BERT',
        "max_seq_length": 512,
        "eval_batch_size": 2,
    }

    import random
    from tqdm import tqdm
    from pytorch_pretrained_bert.tokenization import BertTokenizer
    from modeling_readmission import BertForSequenceClassification
    from torch.utils.data import TensorDataset, SequentialSampler, DataLoader
    from torch.utils.data.distributed import DistributedSampler
    import torch

    processors = {
        "readmission": ReadmissionProcessor
    }

    if config['local_rank'] == -1 or config['no_cuda']:
        device = torch.device("cuda" if torch.cuda.is_available() and not config['no_cuda'] else "cpu")
        n_gpu = torch.cuda.device_count()
    else:
        device = torch.device("cuda", config['local_rank'])
        n_gpu = 1
        # Initializes the distributed backend which will take care of sychronizing nodes/GPUs
        torch.distributed.init_process_group(backend='nccl')
    logger.info("device %s n_gpu %d distributed training %r", device, n_gpu, bool(config['local_rank'] != -1))

    random.seed(config['seed'])
    np.random.seed(config['seed'])
    torch.manual_seed(config['seed'])
    if n_gpu > 0:
        torch.cuda.manual_seed_all(config['seed'])

    if os.path.exists(config['output_dir']):
        pass
    else:
        os.makedirs(config['output_dir'], exist_ok=True)

    task_name = config['task_name'].lower()
```

```python
    if task_name not in processors:
        raise ValueError(f"Task not found: {task_name}")

    processor = processors[task_name]()
    label_list = processor.get_labels()

    tokenizer = BertTokenizer.from_pretrained('bert-base-uncased')

    # Prepare model
    model = BertForSequenceClassification.from_pretrained(config['bert_model'], 1)
    if config['fp16']:
        model.half()
    model.to(device)
    if config['local_rank'] != -1:
        model = torch.nn.parallel.DistributedDataParallel(model, device_ids=[config['local_rank']],output_device=config['local_rank'])
    elif n_gpu > 1:
        model = torch.nn.DataParallel(model)

    eval_examples = processor.get_test_examples(config['data_dir'])
    eval_features = convert_examples_to_features(
        eval_examples, label_list, config['max_seq_length'], tokenizer)
    logger.info("***** Running evaluation *****")
    logger.info("  Num examples = %d", len(eval_examples))
    logger.info("  Batch size = %d", config['eval_batch_size'])
    all_input_ids = torch.tensor([f.input_ids for f in eval_features], dtype=torch.long)
    all_input_mask = torch.tensor([f.input_mask for f in eval_features], dtype=torch.long)
    all_segment_ids = torch.tensor([f.segment_ids for f in eval_features], dtype=torch.long)
    all_label_ids = torch.tensor([f.label_id for f in eval_features], dtype=torch.long)
    eval_data = TensorDataset(all_input_ids, all_input_mask, all_segment_ids, all_label_ids)
    if config['local_rank'] == -1:
        eval_sampler = SequentialSampler(eval_data)
    else:
        eval_sampler = DistributedSampler(eval_data)
    eval_dataloader = DataLoader(eval_data, sampler=eval_sampler, batch_size=config['eval_batch_size'])
    model.eval()
    eval_loss, eval_accuracy = 0, 0
```

```python
    nb_eval_steps, nb_eval_examples = 0, 0
    true_labels = []
    pred_labels = []
    logits_history = []
    m = torch.nn.Sigmoid()
    for input_ids, input_mask, segment_ids, label_ids in tqdm(eval_dataloader):
        input_ids = input_ids.to(device)
        input_mask = input_mask.to(device)
        segment_ids = segment_ids.to(device)
        label_ids = label_ids.to(device)
        with torch.no_grad():
            tmp_eval_loss, temp_logits = model(input_ids, segment_ids, input_mask, label_ids)
            logits = model(input_ids, segment_ids, input_mask)

        logits = torch.squeeze(m(logits)).detach().cpu().numpy()
        label_ids = label_ids.to('cpu').numpy()

        outputs = np.asarray([1 if i else 0 for i in (logits.flatten() >= 0.5)])
        tmp_eval_accuracy = np.sum(outputs == label_ids)

        true_labels = true_labels + label_ids.flatten().tolist()
        pred_labels = pred_labels + outputs.flatten().tolist()
        logits_history = logits_history + logits.flatten().tolist()

        eval_loss += tmp_eval_loss.mean().item()
        eval_accuracy += tmp_eval_accuracy

        nb_eval_examples += input_ids.size(0)
        nb_eval_steps += 1

    # 绘制评价曲线
    df = pd.DataFrame({'logits': logits_history, 'pred_label': pred_labels, 'label': true_labels})
    df_test = pd.read_csv(os.path.join(config['data_dir'], "sample.csv"))

    fig = plt.figure(1)
    ax1 = fig.add_subplot(1,2,1)
    ax2 = fig.add_subplot(1,2,2)
    fpr, tpr, df_out = vote_score(df_test, logits_history, ax1)
    rp80 = vote_pr_curve(df_test, logits_history, ax2)

    output_eval_file = os.path.join(config['output_dir'], "eval_results.txt")
```

```
plt.tight_layout()
plt.show()
```

6. 模型评估

```
eval_loss = eval_loss / nb_eval_steps
eval_accuracy = eval_accuracy / nb_eval_examples
result = {'eval_loss': eval_loss,
          'eval_accuracy': eval_accuracy,
          'RP80': rp80}
with open(output_eval_file, "w") as writer:
    logger.info("***** Eval results *****")
    for key in sorted(result.keys()):
        logger.info("  %s = %s", key, str(result[key]))
        writer.write("%s = %s\n" % (key, str(result[key])))
```

评估结果如图 1.10 所示。

```
02/11/2022 22:31:41 - INFO - __main__ -   ***** Eval results *****
02/11/2022 22:31:41 - INFO - __main__ -     RP80 = 0.5

02/11/2022 22:31:41 - INFO - __main__ -     eval_accuracy = 0.8333333333333334

02/11/2022 22:31:41 - INFO - __main__ -     eval_loss = 0.5382881859938303
```

图1.10　评价结果

由图 1.10 中可以看出，模型较为准确。感兴趣的同学可以调增参数，进一步提升模型精度。

单 元 小 结

在本单元中，同学们学习了自然语言处理的基础知识，了解了相关子任务，也学习了机器学习和自然语言处理如何应用于实际场景，在后续单元将学习自然语言相关任务背后的科学原理。后续单元将介绍机器学习在自然语言处理中的应用，通过机器学习可以避免预测自然语言所有的表达方式（泛化能力），不仅可以用于简单任务完成，还可以帮助实现商业和生活中的目标。

在下一单元中，将介绍分词和字典相关知识，并使用 HanLP 解决一个简单的文本分词任务。

习　题

1. 什么是自然语言处理？
2. 自然语言处理的应用有哪些？

单元2 分词和字典

如果已经准备好用自然语言处理能力来解决实际问题,那么第一件事就是构建自己的词汇表。本单元将帮助同学们将文档或者任何字符串拆分为离散的、有意义的词条并构建字典。

分词是将文本序列划分为词语的过程,虽然看似简单,但该过程是所有自然语言处理下游应用的基础,而字典树是分词的一种有效存储方式。没有一个良好的分词和字典构建方案,所有的自然语言处理应用都只能是空中楼阁。所以,本单元内容非常重要,是后续所有单元的基础,同学们一定要好好掌握。下面将从分词、字典树、切分算法和评测指标四个方面展开。本单元的知识导图如图2.1所示。

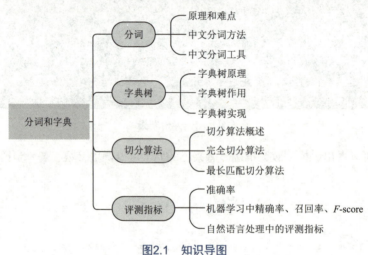

图2.1 知识导图

课程安排

课程任务	课程目标	安排学时
了解分词的基本概念	了解分词的原理和难点,掌握常用的分词方法和分词工具	1
掌握字典树的使用方法	了解字典树的概念及作用,熟练掌握字典树的使用方法	1
了解常用切分算法	了解切分算法的概念,并能根据实际需要选择合适的切分算法	1
掌握评测指标的定义及作用	熟悉常用的评测指标,并能合理运用于自然语言评测中	1
实战应用	通过实战应用引导同学们了解业务背景,解决实际问题,进行自我练习、自我提高	2

2.1 分　　词

2.1.1 中文分词的原理和难点

通俗而言，汉语分词就是在汉语语句的词与词之间直接加上空格或者其他分界符号，也就是将一段文本划分为各种词语。中文分词的方法有很多，可以根据原理分成基于字典的分词方法、基于统计学模型的分词方法、基于规则的分词方法这几个类型，如图2.2所示。本单元将详细介绍基于字典的分词方法。

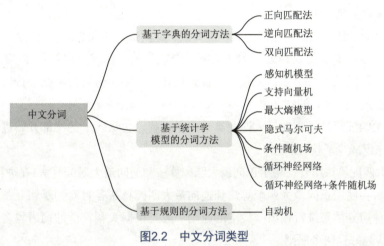

图2.2　中文分词类型

中文分词的难点主要集中在三个方面：分词的规范、歧义词的切分、未登录词的识别。

（1）在分词的规范方面，由于中文自身语言的特性，"词"的概念一直是汉语语言学界挥之不去的问题，究竟词的抽象定义是什么，以及词的具体界定是什么，至今依然是无法得到有效解答的，至今依然没有一个公认的且具有权威性的词表。主要问题出在单字词与词素的划界和词与短语的划界。尽管在1992年国家颁布了《信息处理用现代汉语分词规范》，但是这种规范很容易受主观因素影响，在处理现实问题时效果欠佳。

（2）在歧义词的切分方面，中文中的歧义词是普遍存在的，对同一个中文词语可以有多种切分方式，为了更好地区分这些歧义现象，可以将中文歧义词分为交集型切分歧义、组合型切分歧义、混合型切分歧义这三种类型。

针对ABC类型的汉语词汇，如果AB可以单独作为一个词语，同时BC也可以单独作为一个词语，这种情况就属于交集型切分歧义。例如，词语"研究生命"可以切分为"研究/生命"，也可以切分为"研究生/命"；词语"很热闹"可以同时切分为"很热/闹"和"很/热闹"，很难界定哪一种切分是正确的。组合型切分歧义是指，词语如AB，同时满足A、B、AB都可以单独成词。比如，"他将来我的住处"中的"将来"一词，同样会造成切分方式选择的困扰。换句话说，当一个字符串可以在这里切开也可以在那里切开，这属于交集型歧义；一个字符串可以选择切开也可以选择不切开，这属于组合型歧义。混合型切分歧义是指汉语词语同时包含以上两种情况。

（3）在未登录词（新词）的识别方面，两种情况可以造成未登录词语的出现：一是词库中没有收录这个词语；二是当前用于训练模型的语料库中没有出现这个词语。未登录词语有的是研究领域新出现的专有名词，如"埃博拉""新冠"等；有的是企业名称或城市名称，抑或是电影、书籍名称，如"北京市""长津湖""自然语言处理技术与应用"等；还有的是新出现的网络用语，比如"烟火气""天花板""沉浸式"等。

2.1.2 基于字典的中文分词方法

基于字典的分词方法又可以称为机械切分方法，它的基本原理是：选取句子中相邻的字构成词语，随后去字典中进行比对，如果字典中存在这个词语则切分完成，如果不存在则按照设定的规则重新选字构词，并继续到字典中比对。

基于字典的分词方法可以分为正向最大匹配算法、逆向最大匹配算法、双向最大匹配算法。

（1）正向最大匹配算法会从待处理句子的第一个字开始，在字典中寻找能匹配的最长的词语，如果找不到则会从最后减去一个字，依此类推直到只剩一个字，随后会在此进行切分。正向最大匹配算法的缺点是它会产生很大的歧义。

（2）逆向最大匹配算法是正向最大匹配算法的逆过程，它从句子的尾部开始进行匹配，相对正向最大匹配算法来说减少了歧义。

（3）双向最大匹配算法结合使用了正向最大匹配算法和逆向最大匹配算法，在进行分词的过程中，双向最大匹配算法会使用正向最大匹配算法和逆向最大匹配算法各自对句子切分一遍，然后根据设定的原则选取一种切分结果进行输出。设定的原则一般是根据大颗粒度的词语越多越好和非字典词语和单字词语越少越好这两个原则。

以上三种切分算法的优点是显而易见的：实现简单。其缺点是局限性大，对于常见的一词多义和歧义问题等无法进行有效处理。相对于这些机械切分方法，基于统计学模型的分词方法能更好地处理上述问题。

2.1.3 主流的中文分词工具

表2.1罗列了一些主流的中文分词工具，在实际学习与实践中，要根据实际需要选择最适合的工具。

表2.1 主流的中文分词工具

分词工具	是否开源	简 介
Jieba	开源	优秀的Python中文分词工具，支持三种分词模式：精确模式、全模式和搜索引擎模式
Jiagu	开源	集成多种NLP基础处理功能，并支持知识图谱开放信息抽取
SnowNLP	开源	优秀的Python中文分词工具
HanLP	开源	优秀的工具包，采用全世界量级最大、种类最多的语料库，具备功能完善、性能高效、架构清晰、语料时新、可自定义的特点
pyltp（哈工大语言云）	付费使用	哈工大自然语言工作组推出的基于Python封装的自然语言处理工具
THULAC(清华中文词法分析)	付费使用	清华大学自然语言处理与社会人文计算实验室研制推出的中文词法分析工具包，具有中文分词和词性标注功能

在上述分词工具中，Jieba 分词和 HanLP 分词的安装和使用相对简单，对于初学者相对友好，同时也有着较为强大的功能。下面以 Jieba 分词和 HanLP 分词为例介绍分词工具的基本工作原理。

Jieba 分词原理简介：①根据 Jieba 自带的（或是用户自己加载的）用于统计的词典构造出词的前缀词典；②在词典创建好之后，使用该前缀词典处理分词的句子，经过处理生成一个 DAG 有向无环图；③根据动态规划算法在前一步构造的有向无环图中找出一条最有可能走的路径，也就是最大的概率，以此进行分词。除此之外，如果出现的词不在统计词典中，使用隐马尔可夫模型，配合 Viterbi 维特比算法，可以成功找出可能出现的隐状态序列。

HanLP 分词原理简介：HanLP 中文分词方法有标准分词、NLP 分词、索引分词、N- 最短路分词、CRF 分词以及极速分词等。标准分词（HanLP.segment）又称为最短路分词，采用 Viterbi 维特比算法求解最短路。NLP 分词（NLPTokenizer）会执行命名实体识别和词性标注。索引分词（IndexTokenizer）是面向搜索引擎的分词器，它可以对长的词语进行全切分。N- 最短路分词（NShortSegment）对于命名实体识别能力更强，效果更好，但是它比最短路分词要慢一些。CRF 分词（CRFSegment）又称为条件随机场分词，它基于 CRF 模型和 BEMS 标注训练而成，不支持命名实体识别，一般仅用于新词识别。极速分词（AhoCorasickSegment）是词典最长分词，速度最快，但是精度一般。本单元实践部分将使用标准分词。

2.2 字 典 树

2.2.1 字典树概述

字典树是一种树状的数据结构，它利用了字符串的公共前缀，存储时节约了数据存储的空间，并提高了查询时字符串比较的效率。图 2.3 是一棵字典树，其中虚线框框选的节点是人为规定的目标节点，这个字典中的单词就是从 root 根节点出发到人为规定的目标节点的路径。从 root 到达某一目标节点的路径中的字母连起来就是一个单词。图 2.3 所示字典树包括的单词有 a，abc，bac，bbc，ca 五个。

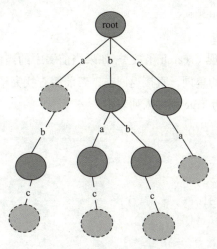

图2.3　字典树示意图

建立一棵字典树之后，就可以利用它完成很多功能，比如，维护字典集合、向字符串集合中插入新的字符串、查询字符串集合中是否含有目标字符串、统计字符串在字符串集合中出现的次数、对字符串集合按照一定规则排序、求字符串集合的最长公共前缀（Longest Common Prefix）。

2.2.2 字典树的作用

在中文分词的过程中，将从待分词的文本中划分出的单词和字典中的已有单词进行比较所花费的时间显著地影响分词的效率。同时，需要考虑字符串的存储所占的空间，不同的数据结构所占的存储空间差距很大，如何才能提高在词典中查询单词的速度并兼顾存储要求呢？工程师们设计了很多的数据结构，其中字典树在效率和内存之间很好地实现了平衡。本单元实战应用所用的 HanLP 工具中，字典树也是使用最多的数据结构。

2.2.3 字典树的简易实现

在本节中将使用 Python 实现一棵简易的字典树，并实现字典树的增加节点和查找功能。

首先定义节点：

```
class tritree:
    def __init__(self):
        self.dicts={}
        self.isWord=False
```

将一个单词加入字典树功能实现（insert 函数）：首先看这个单词当前字母是否在当前节点的字典中，若不在则生成一个节点，让它对应当前字母，即是将当前字母加入当前节点的字典中，然后进入下一个节点，若当前字母在当前节点的字典中，则直接进入下一个节点，当单词遍历完毕，将 isWord 标记置为 True, 表示单词存在。

```
def insert(word, treeNode):
    for s in word:
        if s not in treeNode.dicts:
            treeNode.dicts[s]=tritree()
        treeNode=treeNode.dicts[s]
    treeNode.isWord=True
```

根据前缀查找单词功能实现（search 函数）：如果当前缀中的当前字母不在当前节点的字典中，说明当前字典树不存在该前缀的单词，直接返回空，若前缀查找完毕，那么对于剩下的部分直接进行 dfs 即可，一旦遇到 isWord 为 True 则加入列表中。

```
def dfs(nodes,strs):
    ret=[]
    if nodes.isWord == True:
        ret.append(strs)
    for s,v in nodes.dicts.items():
        tmp=strs
        tmp+=s
```

```
            ret.extend(dfs(nodes.dicts[s],tmp))
    return ret

def serch(prefix, treeNode):
    ret, strs = [], ''
    for s in prefix:
        if s not in treeNode.dicts:
            return []
        strs += s
        treeNode = treeNode.dicts[s]
    ret=dfs(treeNode, strs)
    return ret
```
测试运行效果：
```
obj=tritree()
insert("apple",obj)
insert("app",obj)
print(serch("aps",obj))
insert("apppppple",obj)
print(serch("ap",obj))

运行结果：
#[]
#['app', 'apple', 'apppppple']
```

2.3 切分算法

2.3.1 切分算法概述

确定好字典之后，需要对给定的文本进行分词。对于一个具体的句子，不同的切分算法所得到的分词结果是不同的。比如，句子"矿泉水不如果汁味道好"可以划分为"矿泉水 / 不如 / 果汁 / 味道好"，也可以划分为"矿泉水 / 不 / 如果 / 汁 / 味道好"，这个句子的关键就在于"不如"与"果汁"的正确分词，需要选择合适的切分算法。切分算法主要可以分为基于词表的切分方法、基于统计模型（包括基于 HMM 和 n 元语法）的切分方法。基于字典的切分方法简单好用，是很多分词器都采用的方法，它又可以细分为完全切分算法、最长匹配算法等。本单元将详细介绍基于字典的切分算法。

2.3.2 完全切分算法

完全切分是指找出一段文本中所有的分词。在不考虑效率的情况下，实现完全切分算法非常简单，只需要不断遍历文本中的连续序列，查询当前序列是否在字典中出现即可。完全切分算法的主要问题是：这个算法会把单个字全部输出，并没有考虑到是否是有意义的词语序列。例如，"手机与电视"

完全切分得到的结果是['手',' 手机',' 与',' 电',' 电视',' 视']。这样的分词结果多了很多冗余的词语，通常不是所需的。

完全切分算法代码实现如下所示，只需遍历文本中词是否在词典中即可。

```python
def fully_segment(text, dic):
    word_list = []
    for i in range(len(text)):
        for j in range(i, len(text)+1):
            word = text[i:j]
            if word in dic:
                word_list.append(word)
    return word_list
if __name__ == '__main__':
    dic = load_dictionary()
    # load_dictionary() 需自行定义
    word_list = fully_segment("商品和服务", dic)
    print(word_list)
```

2.3.3 最长匹配算法

最长匹配算法分为正向最长匹配算法、逆向最长匹配算法、双向最长匹配算法等。它们的区别主要在于对待切分的文本的扫描方向不同：如果扫描的顺序从左到右，则称为正向最长匹配算法；如果扫描的顺序从右到左，则称为逆向最长匹配算法；同理，有双向最长匹配算法。相比于完全切分算法，最长匹配算法实际上是做了一个约束，在遍历的过程中优先匹配最长的字符串序列。本节将着重介绍正向最长匹配算法。正向最长匹配算法的基本原理是：从待切分句子的左侧开始，与字典中的字符串进行匹配，以字典中匹配到的最长的字符串作为匹配的结果。如下代码是使用Python实现的正向最长匹配算法代码实例。

```python
def forward_segment(text, dic):
    word_list = []
    i = 0
    while i <len(text):
        longest_word = text[i]
        for j in range(i+1, len(text)+1):
            word = text[i:j]
            if word in dic:
                longest_word = word
        i += len(longest_word)
        word_list.append(longest_word)
    return word_list
```

2.4 评测指标

实现分词任务之后,如果评价分词结果的好坏呢?需要依照什么标准呢?例如,原句为"我喜欢学习自然语言处理",分词方法一的结果是"我 / 喜欢学习 / 自然 / 语言处理",分词方法二的结果是"我 / 喜欢 / 学习 / 自然语言处理",如何给这两种分词结果打分呢?

2.4.1 机器学习中的准确率

准确率(Accuracy)是用来权衡模型预测结果与实际结果差距的指标,准确率越接近 1 说明分类结果越准确。例如,现在有 1 000 条外卖评论文本需要进行分类,需要判断评论是正面还是负面。模型预测的结果 660 段评论是正面的,340 条评论是负面的,真实情况是预测为负面的 340 条评论中有 210 条是分类正确的,660 条预测为正面的评论中有 570 条是分类正确的。可以很自然地想到将准确率设定为(210+570)/1 000=0.78,就是将正确划分的数量除以总的数量,这就是准确率了。

但是,如此好理解的准确率带来了问题,比如,假设又增加了 1 000 条评论数据,真实情况是正面评论 800 条,负面评论 200 条,如果分类用的模型恰好是坏的:它对于任意评论的输出结果都是正面的,那么最后模型的分类结果就是正面的评论全判断对了,负面评论全判断错了,这时候的准确率为(800+0)/1 000=0.8。这种情况下准确率就无法对模型效果做出合理的评估,于是需要用到精确率和召回率。

2.4.2 机器学习中的精确率、召回率与 F-score

精确率(Precision)计算的是所有正确分类的样本与总样本数比例。召回率(Recall)也叫查全率,反映正样本被预测为正的比例。下面举例阐述它们的含义。仍然假设有 1 000 条外卖评论文本需要进行分类,需要判断评论是正面还是负面。

通过表 2.2 可以知道,真实数据集中有 700 条正面评论(570+130),同时有 300 条负面评论(80+220),同时 570 和 130 这两个数据的含义是对于数据集中的 700 条正面评论,模型预测正确(将正面评论预测为正面评论)的有 570 条,预测错误(将正面评论预测为负面评论)的有 130 张,同理 80 和 220 是对于数据集的 300 条负面评论,模型预测正确的有 220 条,预测错误的有 80 条。

表2.2 外卖评论案例混淆矩阵

真 实	预 测	
	正面评论	负面评论
正面评论	570	130
负面评论	80	220

将表 2.2 抽象一下就得到了表 2.3。

表2.3 混淆矩阵

真 实	预 测	
	P	N
P	TP	FN
N	FP	TN

TP（True Positive）：表示将正样本预测为正样本，即预测正确。

FP（False Positive）：表示将负样本预测为正样本，即预测错误。

FN（False Negative）：表示将正样本预测为负样本，即预测错误。

TN（True Negative）：表示将负样本预测为负样本，即预测正确。

将表 2.3 称为混淆矩阵（Confusion Matrix），由此可以定义准确度、精确率、召回率与 F-score。

$$\text{Accuracy} = \frac{TP+TN}{TP+FP+FN+TN}$$

$$\text{Precision} = \frac{TP}{TP+FP}$$

$$\text{Recall} = \frac{TP}{TP+FN}$$

$$F\text{-score} = (1+\beta^2)\frac{\text{Precision}\,\text{Recall}}{\beta^2(\text{Precision}+\text{Recall})}$$

注意：F-score 中 β 是一个数值，通常使用 β 为 1 时的 F-score，也称为 F_1 值。

在上面的外卖评论正面负面分类例子中，对于正面评论有

$$\text{Precision} = \frac{570}{570+80} = 0.88$$

$$\text{Recall} = \frac{570}{570+130} = 0.81$$

$$F_1 = \frac{2\times 0.88\times 0.81}{0.88+0.81} = 0.84$$

通过引入精确率、召回率能够明显地弥补只用准确率的不足，同时加入 F-score 能够弥补召回率和精确率的不足。

2.4.3　NLP 中的精确率、召回率和 F-score

在中文分词中，和上面的评论分类问题不同，标准答案和模型预测的分词结果单词数量不一定一样。前面的评论分类属于分类问题，中文分词属于分块（Chunking）问题。可以做一个思维转换，对于长度为 n 的字符串，使用模型分词的结果是一系列单词。设每个单词按照它在文中的起止位置可以记为区间 $[i,j]$，其中 $1\leqslant i\leqslant j\leqslant n$。那么标准答案所有区间构成的集合 A 作为正类，其他情况作为负类，同时，即分词结果所有单词的区间集合为 B，则有

TP $\cup$ FN=A，TP $\cup$ FP=B，$A\cap B$=TP。由此有

$$\text{Precision} = \frac{|A+B|}{|B|},\quad \text{Recall} = \frac{|A\cap B|}{|A|}$$

例如，对于句子"南京市长江大桥"，见表 2.4。

表2.4 分词案例1

	分词结果	分词区间	
标准答案	['南京市', '长江大桥']	[1,2,3]，[4,5,6,7]	A
分词结果1	['南京', '市长', '江大桥']	[1,2]，[3,4]，[5,6,7]	B
重合部分	无		$A\cap B$
分词结果2	['南京市', '长江大桥']	[1,2,3]，[4,5,6,7]	B'
重合部门	['南京市', '长江大桥']	[1,2,3]，[4,5,6,7]	$A\cap B$

可以发现，重合部分就是正确的部分。对于分词结果1来说，没有重合部分（正确部分），精确度和召回率均为0。对于分词2来说，精确度和召回率都是1。

又如，对于语句"矿泉水不如果汁味道好"，见表2.5。

表2.5 分词案例2

	分词结果	分词区间	
标准答案	['矿泉水', '不如', '果汁', '味道', '好']	[1,3]，[4,5]，[6,7]，[8,9]，[10,10]	A
分词结果1	['矿泉水', '不', '如果','汁', '味道', '好']	[1,3]，[4,4]，[5,6]，[7,7]，[8,9]，[10,10]]	B
重合部分	['矿泉水', '味道', '好']	[1,3]，[8,9]．[10,10]	$A\cap B$

这个例子的精确度为 3/6=0.5，召回率为 3/5=0.6。同理可以得到 F_1=0.545。

实战应用：使用HanLP词典实现中文分词

应用背景：学计算机技术的你毕业后进入了一家网站广告设计公司，在产品推介时，客户对于公司的产品提出了改进建议，希望网站的搜索功能对于关键词的定位更加准确、覆盖面更广。产品经理在设计页面标题的时候，基本原则是覆盖相关的关键词，如果想要覆盖的词有很多个，并且只是单纯地全部列举出来，这不仅会影响到用户体验，同时也可能触犯搜索引擎规则。而公司希望 AI 技术团队能够尽快解决该问题。

这一问题可以通过应用中文分词技术解决，可以达到了覆盖目标关键词的目的。本次实战推荐使用 HanLP。HanLP 的使用相对简单，非常适合新手使用。本节使用 Python 语言实现，在进行下面的操作之前，要确保操作环境中 Python 版本为 3.6 以上，且由于 HanLP 主项目采用 Java 开发，所以还需要 Java 运行环境，需提前安装 JDK。

（一）安装 HanLP

```
pip install pyhanlp
```

注意：若未安装 Java 则会报如下错误，则应安装 JDK。

```
jpype.jvmfinder.JVMNotFoundException: No JVM shared library file (jvm.dll) found. Try setting up the JAVAHOME environment variable properly.
```

(二)分词实战

简单样例：输入如下代码并运行，对 text 的内容进行标准分词并打印结果。

```
import pyhanlp
text = '铁甲网是中国最大的工程机械交易平台'
words = []
for term in pyhanlp.HanLP.segment(text):
    words.append(term.word)
print(words)    #输出分词结果
```

运行结果如下：

```
['铁甲', '网', '是', '中国', '最大', '的', '工程', '机械', '交易', '平台']
```

自定义分词词典：可以通过自定义增加不同领域的专业名词词典，从而使得 HanLP 分词的结果更加符合需要。在之前代码基础上继续输入如下代码：

```
CustomDictionary.add("铁甲网")
    CustomDictionary.insert("工程机械", "nz 1024")
    CustomDictionary.add("交易平台", "nz 1024 n 1")
    print(HanLP.segment(txt))
```

运行结果如下：

```
['铁甲网', '是', '中国', '最大', '的', '工程机械', '交易平台']
```

单 元 小 结

分词技术属于自然语言处理的子任务，它是机器翻译、文本搜索、语音合成、自动分类等其他信息处理任务的基础。本单元首先介绍了汉语分词的原理及难点，并介绍了汉语分词常用的开源工具；然后引入了用于建立分词用字典的数据结构字典树，在得到字典之后，又引入了两种经典的单词切分算法：完全切分算法和最长匹配算法，并给出了它们的代码实现；随后，本单元对一些常用的评价指标进行了介绍，包括准确率、精确率、召回率和 F-score；最终，以一个用 HanLP 实现中文分词的实践结尾。本单元对分词进行了详细介绍，从原理到应用逐渐深入，力求让同学们快速掌握其思想及原理。

下一单元将介绍自然语言处理中必不可少、至关重要的步骤——数据预处理。

习 题

1. 什么是自然语言处理中的分词？
2. 什么是自然语言处理中的字典？

单元3 数据预处理

在上一单元中介绍了如何进行分词并采用合理的数据结构进行存储。在现实世界中,所使用的数据并不是完美的,它们有的数据缺失,有的具有异常值和嘈杂值,这时就需要对其进行数据预处理,将其变成完美的数据并将其送入机器学习算法中进行运算。

数据预处理在自然语言处理中是一项关键步骤,它常常可以决定一个项目的成败。本单元将介绍数据预处理的基本步骤,要求同学们了解并能够运用数据预处理的多种方式。本单元知识导图如图3.1所示。

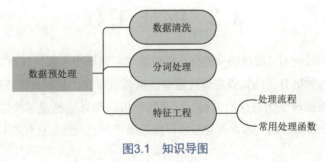

图3.1 知识导图

课程安排

课 程 任 务	课 程 目 标	安排学时
了解数据清洗	了解数据清洗的概念及作用	1
了解分词处理	了解分词处理的概念及方法	1
掌握特征工程方法	掌握特征工程处理流程,并能根据任务需求选择合适的处理方法	1
实战应用	通过该实战应用引导同学们了解一些业务背景,解决实际问题,进行自我练习、自我提高	2

3.1 数据清洗

数据清洗,顾名思义就是将要用到的数据中重复、多余部分的数据进行筛选并清除,把缺失部分补充完整,并将不正确的数据纠正或者删除,最后整理成可以进一步加工、使用的数据。

数据清洗是为了洗掉数据集中的脏数据。脏数据一般为数据集中残缺、错误、重复的数据。数据清洗是为了提高数据的质量、缩小数据统计过程中的误差值。

不同的异常数据所使用的数据清洗的方法也不一样。

重复数据常用删除法；缺失值数据常用删除法、替换法、插补法；异常值数据常用删除法、替换法，不敏感的可不做处理，部分情况下也可以视作缺失值数据处理。

3.2 分词处理

在进行文本挖掘的时候，最先进行的数据预处理步骤就是分词。英文单词在拼写时会有空格将其隔开，可以很容易地按照空格来进行分词，但是也有不少词语由多个单词组成，比如一些专有名词，如"Thanksgiving Day"。而中文并没有空格来辅助分词，因此分词就需要专门进行解决。

对于文本挖掘中所需要实现的分词部分，通常会使用已有工具。简单的英文文本进行分词不需要借助任何工具，通过空格和标点符号就可以，而进一步的准确进行英文分词则推荐使用 nltk。而对于中文分词，推荐用 Jieba 分词。

3.3 特征工程

特征工程是机器学习中不可或缺的一部分，在机器学习领域中占据极其重要的位置。

特征工程是利用与数据有关的知识来设计能够使机器学习算法达到最优特征的过程。通俗地说，特征工程就是将原始数据转变为特征的过程，这些特征可以很好地描述这些数据，而且利用这些特征建立的模型在大部分未知数据上的表现甚至可以接近最佳乃至达到最优性能。

3.3.1 处理流程

特征工程的主要有四个步骤，分别为：

（1）数据清洗：处理数据的不一致性，如重复值、异常值、缺失值处理。

（2）数据集成：将多个数据源汇合并统一存储，如建立数据库。

（3）数据变换：结合分析需求，对数据进行标准化、离散化、稀疏化等处理。

（4）数据归约：在保持数据完整性的基础上，有效降低数据规模，得到数据集的归约表示。

3.3.2 常用的中文文本处理函数

为了在学习工作中使用方便，本单元对常见的中文文本处理函数进行了总结。

1. 数据获取

使用公开的语料库或者网络爬虫获取，常用的数据获取函数如下：

```
# 读取文件列表数据，返回文本数据的内容列表和标签列表
def filelist_contents_labels(filelist):
    contents = []
    labels = []
    for file in filelist:
```

```
            with open(file, "r", encoding="utf-8") as f:
                for row in f.read().splitlines():
                    sentence=row.split('\t')
                    contents.append(sentence[-1])
                    if sentence[0]=='other' :
                        labels.append(0)
                    else:
                        labels.append(1)
        return contents,labels
```

2. 文件读写

（1）CSV 文件读取

```
import csv
with open(path,"r",encoding="utf-8") as f:
    reader=csv.reader(f) #csv 阅读器 默认分隔符为",", 设置分隔符用 delimiter
    #reader=csv.reader(f,delimiter="\t")
    birth_header=next(reader) # 获取首行标签
    for row in reader: # 遍历文本
        ...
```

（2）CSV 文件写入

```
import csv
f=open("data.csv","w",encoding="utf-8",newline="")
f_writer=csv.writer(f,delimiter=" ") # delimiter 记录分隔符
f_writer.writerow(["label","text"]) # 列表形式，写入首行标签
for line in lines:
    f_writer.writerow(line) # 写入数据
        ...
f.close()
```

（3）Json 文件读取

```
import jsonlines
with open("data.jsonl","r") as f:
    lines=jsonlines.Reader(f)
    for line in lines: # line 为字典格式
        ...
# 或者采用以下方法
import json
with open("data.jsonl","r") as f:
    lines=f.readlines()
    for line in lines:
        line=json.loads(line) # 将数据转换为字典格式
        ...
```

（4）Json 文件写入

```
import jsonlines
with open("data.jsonl","w",encoding="utf-8") as f:
    pass
with jsonlines.open("data.jsonl",mode="a") as f:
    f.write(input_line) # input_line 为字典格式
        ...
# 或者采用以下方法
import json
with open("data.jsonl","w",encoding="utf-8") as f:
    f.write(json.dumps(input_line)+"\n") # input_line 为字典格式
        ...
```

3. 格式转换

（1）全角、半角转换。在自然语言处理过程中，全角、半角的不一致会导致信息抽取不一致，因此需要统一。

```
    # 全角转半角
    def full_to_half(sentence):              # 输入为一个句子
        change_sentence=""
        for word in sentence:
            inside_code=ord(word)
            if inside_code==12288:           # 全角空格直接转换
                inside_code=32
            elif inside_code>=65281 and inside_code<=65374: # 全角字符（除空格）根据
关系转化
                inside_code-=65248
            change_sentence+=chr(inside_code)
        return change_sentence

    # 半角转全角
    def hulf_to_full(sentence):              # 输入为一个句子
        change_sentence=""
        for word in sentence:
            inside_code=ord(word)
            if inside_code==32:              # 半角空格直接转换
                inside_code=12288
            elif inside_code>=32 and inside_code<=126:     # 半角字符（除空格）根据
关系转化
                inside_code+=65248
            change_sentence+=chr(inside_code)
        return change_sentence
```

（2）大写数字转小写数字。

```
    # 大写数字转换为小写数字
```

```python
def big2small_num(sentence):
    numlist = {"一":"1","二":"2","三":"3","四":"4","五":"5","六":"6","七":"7","八":"8","九":"9","零":"0"}
    for item in numlist:
        sentence = sentence.replace(item, numlist[item])
    return sentence
```

（3）大写字母转小写字母。

```python
# 大写字母转为小写字母
def upper2lower(sentence):
    new_sentence=sentence.lower()
    return new_sentence
```

（4）去除表情符号。

```python
# 去除文本中的表情字符（只保留中英文和数字）
def clear_character(sentence):
    pattern1= '\[.*?\]'
    pattern2 = re.compile('[^\u4e00-\u9fa5^a-z^A-Z^0-9]')
    line1=re.sub(pattern1,'',sentence)
    line2=re.sub(pattern2,'',line1)
    new_sentence=''.join(line2.split()) # 去除空白
    return new_sentence
```

（5）去除所有符号。

```python
# 去除字母数字表情和其他字符
def clear_character(sentence):
    pattern1='[a-zA-Z0-9]'
    pattern2 = '\[.*?\]'
    pattern3 = re.compile(u'[^\s1234567890::' + '\u4e00-\u9fa5]+')
    pattern4='[\'!"#$%&\'()*+,-./:;<=>?@[\\]^_`{|}~]+'
    line1=re.sub(pattern1,'',sentence)      # 去除英文字母和数字
    line2=re.sub(pattern2,'',line1)         # 去除表情
    line3=re.sub(pattern3,'',line2)         # 去除其他字符
    line4=re.sub(pattern4, '', line3)       # 去掉残留的冒号及其他符号
    new_sentence=''.join(line4.split())     # 去除空白
    return new_sentence
```

（6）简体与繁体转换。

```python
from langconv import *
# 转为简体
def Traditional2Simplified(sentence):
    sentence = Converter('zh-hans').convert(sentence)
    return sentence
# 转为繁体
def Simplified2Traditional(sentence):
```

```
        sentence = Converter('zh-hant').convert(sentence)
        return sentence
```

(7)停用词过滤。

```
# 去除停用词，返回去除停用词后的文本列表
def clean_stopwords(contents):
    contents_list=[]
    stopwords = {}.fromkeys([line.rstrip() for line in open('data/stopwords.txt', encoding="utf-8")])  # 读取停用词表
    stopwords_list = set(stopwords)
    for row in contents:          # 循环去除停用词
        words_list = jieba.lcut(row)
        words = [w for w in words_list if w not in stopwords_list]
        sentence=''.join(words)    # 去除停用词后组成新的句子
        contents_list.append(sentence)
    return contents_list
```

(8)字符串转为列表。

```
import ast
with open("data.txt","r",encoding="utf-8") as f:
    lines=f.readlines()
    for line in lines: #line: "["id","text","label"]"
        line=ast.literal_eval(line) #line: ["id","text","label"]
        img=line[0]
        ...
```

(9)文本填充。

```
    def fill_sentence(embeddings, embedding_dim):
"""
输入:
embeddings: [sentence.array,sentence.array,...]
embedding_dim: 与embeddings中一致
"""
        fill_embeddings = []
        
        length = [len(embedding) for embedding in embeddings] # 一个embedding为一个sentence
        max_len = max(length)
        
        for embedding in embeddings:
            if len(embedding) < max_len:
                fill_zero = np.zeros((max_len - len(embedding), embedding_dim))
                fill_embedding = np.append(embedding, fill_zero)
                fill_embedding = fill_embedding.reshape(-1, embedding_dim)
# -1: reshape函数根据另一个参数的维度计算出数组的另外一个shape属性值
```

```
        fill_embeddings.append(fill_embedding)
    else:
        fill_embeddings.append(embedding)
return np.array(fill_embeddings)
```

实战应用一：英文新闻资讯数据清洗

实战背景：

学习信息技术的你毕业后进入了一家新媒体公司的信息技术部，所在部门主要负责利用自然语言处理方法对新闻资讯进行分类、筛选以及关键信息提取等工作。为了提升处理效率，团队拟对现有的新闻资讯数据进行清洗，请协助团队成员完成新闻资讯清洗工作。

提示：由于英文和中文语言处理方式不同，建议分别对英文新闻数据和中文信息数据进行清洗。

实践内容：以 AG_News 为数据集进行数据清洗。

实践环境：Python 3.9。

实践代码：

```python
import pandas as pd
from nltk.corpus import stopwords
from nltk.stem import PorterStemmer
from textblob import Word
import re
# 读取数据
data = pd.read_csv('AG_News.csv',encoding ="ISO-8859-1")
data.columns
```

```
Index(['V1', 'V2', 'V3'], dtype='object')
```

```python
# 查看前 5 行数据
data.head()
```

显示结果如图 3.2 所示。

V1	V2	V3
3	3 Wall St. Bears Claw Back Into the Black (Reuters)	Reuters - Short-sellers, Wall Street's dwindli...
3	3 Carlyle Looks Toward Commercial Aerospace (Reu...	Reuters - Private investment firm Carlyle Grou...
3	3 Oil and Economy Cloud Stocks' Outlook (Reuters)	Reuters - Soaring crude prices plus worries\ab...
3	3 Iraq Halts Oil Exports from Main Southern Pipe...	Reuters - Authorities have halted oil export\f...
3	3 Oil prices soar to all-time record, posing new...	AFP - Tearaway world oil prices, toppling reco...

图3.2　数据格式

```python
# 去除无用数据，第一列是无用数据
data = data[['V2','V3']]
data.head()
```

显示结果如图3.3所示。

V2	V3
3 Wall St. Bears Claw Back Into the Black (Reuters)	Reuters-Short-sellers, Wall Street's dwindli…
3 Carlyle Looks Toward Commercial Aerospace (Reu…	Reuters-Private investment firm Carlyle Grou…
3 Oil and Economy Cloud Stocks' Outlook (Reuters)	Reuters -Soaring crude prices plus worries\ab…
3 Iraq Halts Oil Exports from Main Southern Pipe…	Reuters-Authorities have halted oil export\f…
3 Oil prices soar to all-time record, posing new…	AFP-Tearaway world oil prices, toppling reco…

图3.3 处理结果

```
# 修改表头信息
data = data.rename(columns={"V2":"title","V3":"text"})
data.head()
```

显示结果如图3.4所示。

Title	text
3 Wall St. Bears Claw Back Into the Black (Reuters)	Reuters-Short-sellers, Wall Street's dwindli…
3 Carly le Looks Toward Commercial Aerospace (Reu…	Reuters-Private investment firm Carlyle Grou…
3 Oil and Economy Cloud Stocks' Outlook (Reuters)	Reuters-Soaring crude prices plus worries\ab…
3 Iraq Halts Oil Exports from Main Southern Pipe…	Reuters-Authorities have halted oil export\f…
3 Oil prices soar to all-time record, posing new…	AFP-Tearaway world oil prices, toppling reco…

图3.4 修改表头后的效果

```
# 去除标点符号及两个以上的空格
data['text'] = data['text'] .apply(lambda x:re.sub('[!@#$:) .;,?&]',' ', x.lower ()))
data['text'] = data['text'].apply(lambda x:re.sub(' ','', x))
data ['text'][0]
# 单词转换为小写
data['text'] = data['text'].apply(lambda x:" ".join(x.lower() for x in x.split()))
data['text'][0]
# 去除停止词，如a、an、the、高频介词、连词、代词等
stop=stopwords.words('english')
data['text'] = data['text'].apply(lambda x: " ".join(x for x in x.split() if x not in stop))
data['text'][0]
# 分词处理，希望能够实现还原英文单词原型
st=PorterStemmer()
data['text'] = data['text'].apply(lambda x: " ".join([st.stem(word) for word in x.split()]))
```

```
data['text'] = data['text'].apply(lambda x: " ".join([Word(word).lemmatize() for word in x.split()]))
data['text'][0]
data.head()
```

实战应用二：中文新闻资讯数据清洗

实战背景：同"实战应用一"。

1. 数据集

可以从新闻网站摘取一段新闻作为实践数据集，文本保存格式设置为utf-8。

2. 数据导入

```
# 数据导入
def read_txt (filepath):
    file = open(filepath,'r',encoding='utf-8')
    txt = file.read()
    return txt
```

3. 数据清洗

去除数字、英文字符、标点等非常规字符。

```
# 匹配 [^\u4e00-\u9fa5]
def find_chinese (file):
    pattern = re.compile(r'[^\u4e00-\u9fa5]')
    chinese_txt = re.sub(pattern,'',file)
    return chinese_txt
```

处理结果如图 3.5 所示。

纯中文文本：腾讯体育月日讯史蒂芬库里时隔天后复出勇士不敌猛龙猛龙本场比赛过后取得胜负战绩锁定季后赛成文本赛季联盟第支锁定季后赛的球队是第支是雄鹿比赛开始后库里

图3.5 处理结果

4. 分词

```
import jieba

txt = '''
腾讯体育 3 月 6 日讯  史蒂芬-库里时隔127 天后复出，勇士113-121 不敌猛龙。猛龙本场比赛过后，取得44 胜18 负战绩，锁定季后赛，成为本赛季联盟第 2 支锁定季后赛的球队，第 1 支是雄鹿。
比赛开始后，库里持球组织进攻，明显改变了猛龙的防守，给克里斯和维金斯创造了轻松得分的机会。但在第一节还剩 6 分 11 秒下场时，库里没有得分，2 次三分出手全部偏出。
但在第二节比赛重新登场后，我们看到了那个熟悉的库里。他接球投三分命中，迎着防守人超远压哨三分命中，第三节还迎着洛瑞完成 3+1。那个三分之王和 2 次常规赛 MVP 风采依旧。
'''
```

```
# 全模式
jieba_list = jieba.cut(txt,cut_all=True)
jieba_txt1 = ' '.join(jieba_list)
print('全模式分词:',jieba_txt1)

# 精准模式
jieba_list = jieba.cut(txt, cut_all=False)
jieba_txt2 = ' '.join(jieba_list)
print('精准模式分词:',jieba_txt2)
```

分词结果如图 3.6 所示。

图3.6 分词结果

5. 停用词去除

本实战中选用"哈工大停词表"去除连词、虚词、语气词等无意义词语。

```
# 去除停用词
def seg_sentence(list_txt):
    # 读取停用词表
    stopwords = stopwords = read_txt('哈工大停用词表.txt')
    seg_txt = [ w for w in list_txt if w not in stopwords]
    return seg_txt
```

6. 词频统计

找出对文本影响最重要的词汇,辅助后续模型选型。

```
# 词频统计
def counter(txt):
    seg_list = txt
    c = Counter()
    for w in seg_list:
        if w is not ' ':
            c[w] += 1
    return c
```

7. 特征选择

```
# TF_IDF 计算
def tf_idf(txt):
    corpus_txt = [' '.join(txt)]
    stopword_list = read_txt(r'哈工大停用词表.txt').splitlines()
```

```
        vector = TfidfVectorizer(stop_words=stopword_list)
        tfidf = vector.fit_transform(corpus_txt)
        print(tfidf)
        # 获取词袋模型中的所有词
        wordlist = vector.get_feature_names()
        # tf-idf 矩阵 元素a[i][j]表示j词在i类文本中的tf-idf权重
        weightlist = tfidf.toarray()
        # 打印每类文本的tf-idf词语权重,第一个for遍历所有文本,第二个for便利某一类文本下
的词语权重
        for i in range(len(weightlist)):
            print("-------第 ", i, "段文本的词语tf-idf权重------")
            for j in range(len(wordlist)):
                print(wordlist[j], weightlist[i][j]  )
```

特征选择结果如图3.7所示。

词语	权重
昔日	0.012180638428352398
是否	0.024361276856704795
是因为	0.012180638428352398
显得	0.024361276856704795
普遍认为	0.012180638428352398
暂时	0.012180638428352398
更为	0.012180638428352398
更是	0.012180638428352398
更深	0.012180638428352398
替补	0.024361276856704795
最快	0.012180638428352398
有人	0.012180638428352398
有数	0.012180638428352398
有望	0.012180638428352398
有过	0.012180638428352398
期待	0.012180638428352398
本场	0.012180638428352398
本来	0.012180638428352398

图3.7 特征选择结果

完整的代码如下,同学们可以在自己的数据上测试验证。

```
import nltk
import jieba
```

```python
import re
from collections import Counter
from sklearn.feature_extraction.text import TfidfVectorizer

# 创建去除非中文字符的函数
# 数据清洗,去除标点符号、数字等其他非中文字符
# 匹配 [^\u4e00-\u9fa5]
def find_chinese (file):
    pattern = re.compile(r'[^\u4e00-\u9fa5]')
    chinese_txt = re.sub(pattern,'',file)
    return chinese_txt

# 文件读取
def read_txt (filepath):
    file = open(filepath,'r',encoding='utf-8')
    txt = file.read()
    return txt

# 中文分词
def cut_word(text):
    # 精准模式
    jieba_list = jieba.cut(text, cut_all=False)
    return jieba_list

# 去除停用词
def seg_sentence(list_txt):
    # 读取停用词表
    stopwords = stopwords = read_txt('哈工大停用词表.txt')
    seg_txt = [ w for w in list_txt if w not in stopwords]
    return seg_txt

# 词频统计
def counter(txt):
    seg_list = txt
    c = Counter()
    for w in seg_list:
        if w is not ' ':
            c[w] += 1
    return c
# TF_IDF 计算
def tf_idf(txt):
    corpus_txt = [' '.join(txt)]
    stopword_list = read_txt(r'哈工大停用词表.txt').splitlines()
```

```python
        vector = TfidfVectorizer(stop_words=stopword_list)
        tfidf = vector.fit_transform(corpus_txt)
        print(tfidf)
        # 获取词袋模型中的所有词
        wordlist = vector.get_feature_names()
        # tf-idf 矩阵 元素 a[i][j] 表示 j 词在 i 类文本中的 tf-idf 权重
        weightlist = tfidf.toarray()
        # 打印每类文本的 tf-idf 词语权重，第一个 for 遍历所有文本，第二个 for 便利某一类文本下的词语权重
        for i in range(len(weightlist)):
            print("-------第 ", i, "段文本的词语 tf-idf 权重------")
            for j in range(len(wordlist)):
                print(wordlist[j], weightlist[i][j]   )


# 主函数
if __name__ == "__main__":
    # 读取文本信息
    news = read_txt('新闻（中文）.txt')
    print("原文:",news)
    # 清洗数据，去除无关标点
    chinese_news = find_chinese(news)
    print("原文文本:",news)
    print("纯中文文本:",chinese_news)
    # 结巴分词
    chinese_cut = cut_word(chinese_news)
    print(chinese_cut)
    # 停用词去除
    chinese_sentence = seg_sentence(chinese_cut)
    print(chinese_sentence)
    # 词频统计
    lists = counter(chinese_sentence)
    print(lists)
    for list in lists.most_common(20):
        print(list)
    # TF-IDF 权重计算
    tf_idf(chinese_sentence)
```

单元小结

数据预处理在自然语言处理工作中极为重要，而且自然语言处理中特征构造是否良好，很大程度上取决于所构造的特征数据集的数据特性与文本内容语义吻合高低，因此需要有针对性地根据任

务需求进行数据清洗。

本单元主要通过不同的脏数据来说明数据预处理的各种方式，并介绍数据预处理的概念和数据预处理的各个步骤。要求同学们能够掌握并运用数据预处理的方式，并能够在实战中进行新闻资讯数据清洗。

下一单元将开始学习利用自然语言处理模型进行文本处理，并将学习隐马尔可夫模型。

习 题

1. 什么是自然语言处理中的数据预处理？
2. 自然语言处理中的数据预处理有哪些常见的技术方法？

单元4 语言模型和算法流程

在上一单元中通过数据预处理获取了"干净"的数据,接下来将利用清洗后的数据完成自然语言处理任务,在此过程中,将学习自然语言处理中的经典模型——隐马尔可夫模型(Indemarkov)。隐马尔可夫模型是传统机器学习中的经典模型,在运用于解决自然语言处理任务时表现出了强大的功能,如词类标注、中文分词、文本生成等。本单元知识导图如图4.1所示。

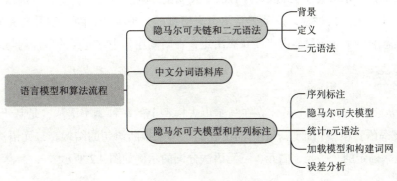

图4.1 知识导图

课程安排

课程任务	课程目标	安排学时
了解隐马尔可夫链	了解马尔可夫背景及定义,以及二元语法的应用	1
了解中文语料库	了解常用的中文语料库,并能够根据实际需要选择合适的语料库	1
掌握隐马尔可夫模型	掌握隐马尔可夫模型的构建方式	1
实战应用	通过该实战应用来引导同学们了解一些业务背景,解决实际问题,进行自我练习、自我提高	2

4.1 隐马尔可夫链和二元语法

隐马尔可夫模型是计算机领域的一个经典假设统计模型,使用范围广泛,接受度高。它在自然语言处理领域用处也非常大,在语音识别、文字识别领域都占据主流地位,这里主要讲解它在中文分词领域的具体使用。隐马尔可夫模型在分词时通常都是与n元语法结合使用,本单元将对隐马尔可夫链和n元语法原理进行讲解。

4.1.1 隐马尔可夫链背景引入

假定某种树的年轮形状有圆形、三角形、正方形、五边形四种,且年轮形状仅受当年的降水状况影响,降水状况有(干旱,丰沛,正常)三种,并且每年的降水情况也仅受到去年降水状况的影响。现在,可以直接观测到很多年前的某树桩的年轮由内向外形状是(三角形、圆形、正方形、五边形、五边形、圆形),这一串形状就属于可见状态链。但是在隐马尔可夫模型中,不仅仅有一串可见状态链,还有一串隐含状态链。在这个例子里,这串隐含状态链就是那几年的降水状况序列,例如,那几年的降水状态可能为(干旱、正常、丰沛、丰沛、丰沛、正常),而类似降水状况的隐含状态链就是隐马尔可夫链。

4.1.2 隐马尔可夫链定义

隐马尔可夫链是马尔可夫(Andrey Markov)发明的一种随机过程,它表示状态空间从一个状态到另一个状态的转换的随机过程,是基于马尔可夫模型的一种统计。隐马尔可夫链是一个无记忆的过程,即下一状态的概率分布只能由当前状态决定,在时间序列中它前面的事件均与之无关。

4.1.3 二元语法

n 元语法(n-gram grammar)是建立在马尔可夫模型上的一种概率语法,是基于(n-1)阶马尔可夫链的一种概率语言模型,通过 n 个语词出现的概率来推断语句的结构。二元语法就是每个字依据前一个字的概率确定的。一元、二元、三元语法分词的示例如图 4.2 所示。

一元语法分词:我、爱、自、然、语、言、处、理

二元语法分词:我爱、爱自、自然、然语、语言、言处、处理

三元语法分词:我爱自、爱自然、自然语、然语言、语言处、言处理

图4.2 语法分词的示例

在中文的自然语言处理中,一般都使用二元语法,因为中文大多数是二字词语。

4.2 中文分词语料库

语料库是自然语言处理中对模型进行训练和检测的文字来源。中文分词利用模型进行分词一般

需要已有的分词语料库对模型进行训练，多根据所需语境使用相应的语料库进行，本书使用的是人民日报分词语料库，每一行都是一句已经分好词的中文语句，大体如图4.3所示。

```
上海  浦东  开发  与  法制  建设  同步
新华社  上海  二月  十日  电
上海  浦东  今年  来  颁布  实行了  涉及  经济  贸易  建设
```

图4.3　人民日报分词语料库示例

4.3　隐马尔可夫模型和序列标注

4.3.1　序列标注

序列标注指的是给定一个序列 $X=\{x_1, x_2\cdots, x_n\}$，找出序列中每个元素对应标签 $y=\{y_1, y_2\cdots, y_n\}$ 的问题。例如，诗句平仄就是一个序列标注。

原始序列：床前明月光，疑是地上霜。

平仄标注：平平平仄平，平仄仄仄平。

对中文分词进行序列标注主要是为了更好地对汉字位置进行捕捉，人们提出了标注集 {B, M, E, S}（其中词首 Begin、词尾 End、词中 Middle、单字成词 Single），具体示例如图4.4所示。

图4.4　序列标注示例

4.3.2　隐马尔可夫模型

隐马尔可夫模型是基于隐马尔可夫链提出的，主要由三个概率矩阵组成：状态转移概率矩阵（在背景引入中就是降水状况互相影响的概率）、发射概率矩阵（在背景引入中为不同降水状况导致不同年轮形状的概率）、初始状态分布（在第一年时的天气状况）。隐马尔可夫模型用于中文分词时，隐马尔可夫链对应的是序列标注，如同背景引入中的天气情况；可见序列对应的是原中文语句，如同背景引入中树桩的年轮情况；状态转移概率矩阵为 BMES 四个状态互相转换的概率矩阵，发射矩阵为 BMES 四个状态分别可能的字的概率矩阵，初始矩阵就是句首第一个字为 BMES 中某一个的概率矩阵。

4.3.3　统计 n 元语法

隐马尔可夫模型进行中文分词第一步就是需要得到三个概率矩阵，也就是统计训练集的二元语法。

```
array_Pi[key]=log(array_Pi[key]/line_num)
```

用于统计初始状态分布，**array_Pi** 为初始的状态矩阵。

```
array_A[key0][key]=log(array_A[key0][key]/count_dic[key0])
```

用于统计状态转移概率矩阵，**array_A** 为状态转移矩阵。

```
array_B[key][word]=log(array_B[key][word]/count_dic[key])
```

用于统计发射概率矩阵，**array_B** 为发射概率矩阵。

4.3.4 加载模型和构建词网

计算好隐马尔可夫模型的三个概率矩阵后，就可以将它们带入模型进行计算。对测试集进行分词主要是动态规划中的维特比算法，目的是通过极大似然估计，寻找最有可能产生观测事件序列的维特比路径，也就是隐含状态序列，原理是某最短路径必然在前一阶段也是最短路径。

利用这些来进行中文分词就是已知可观察到的中文序列，以及隐马尔可夫模型来得到这句话的 BMES 标注序列，遍历每个字，边计算边回溯最可能是答案的路径，在最后剩下的路径中挑选最优路径。公式如下：

```
prob=tab[i-1][state]+array_a[state1][state0]+array_b[state0][sentence[i]]
```

4.3.5 误差分析

一般而言，对分词的误差分析采用召回率和查准率两种。其中某算法的查准率是指通过算法得出的正确分词与算法得出的所有分词的比率；召回率是指通过算法得出的正确分词和所有正确分词的比率。

误差计算示例如图 4.5 所示。

	分词结果	分词区间
标准答案	['中国'，'法律'，'顾问网'，'开通']	[(1,2), (3,4), (5,7), (8,9)]
分词结果	['中国'，'法律'，'顾'，'问网'，'开通']	[(1,2), (3,4), (5,5), (6,7), (8,9)]
重合部分	['中国'，'法律'，'开通']	[(1,2), (3,4), (8,9)]
召回率: 3/4 = 0.75		
查准率: 3/5 = 0.6		

图4.5 误差计算

结合所学内容，下面介绍如何使用隐马尔可夫模型实现中文分词。

```python
from numpy import *
import numpy as np
import sys
import re
STATES = ['B', 'M', 'E', 'S']
array_A = {}            # 状态转移概率矩阵
array_B = {}            # 发射概率矩阵
array_E = {}            # 测试集存在的字符，但在训练集中不存在，发射概率矩阵
```

```python
array_Pi = {}              # 初始状态分布
word_set = set()           # 训练数据集中所有字的集合
count_dic = {}             # 'B,M,E,S' 每个状态在训练集中出现的次数
line_num = 0               # 训练集语句数量

def to_region(segmentation: str) -> list:
    region = []
    start = 0
    for word in re.compile("\\s+").split(segmentation.strip()):
        end = start + len(word)
        region.append((start, end))
        start = end
    return region

def prf(gold: str, pred: str, dic) -> tuple:
    """
    计算P、R、F1
    :param gold: 标准答案文件，比如"商品 和 服务"
    :param pred: 分词结果文件，比如"商品 和服 务"
    :param dic: 词典
    :return: (P, R, F1, OOV_R, IV_R)
    """
    A_size, B_size, A_cap_B_size, OOV, IV, OOV_R, IV_R = 0, 0, 0, 0, 0, 0, 0
    A, B = set(to_region(gold)), set(to_region(pred))
    A_size += len(A)
    B_size += len(B)
    A_cap_B_size += len(A & B)
    # text = re.sub("\\s+", "", gold)
    #
    # for (start, end) in A:
    #     word = text[start: end]
    #     if word in dic:
    #         IV += 1
    #     else:
    #         OOV += 1
    #
    # for (start, end) in A & B:
    #     word = text[start: end]
    #     if word in dic:
    #         IV_R += 1
    #     else:
    #         OOV_R += 1
    p, r = A_cap_B_size / B_size * 100, A_cap_B_size / A_size * 100
```

```python
            if p == 0 :
                f = 0
            else :
                f = 2 * p * r / ((p + r)*1.0)
        return p, r,f
# 初始化所有概率矩阵
def Init_Array():
    for state0 in STATES:
        array_A[state0] = {}
        for state1 in STATES:
            array_A[state0][state1] = 0.0
    for state in STATES:
        array_Pi[state] = 0.0
        array_B[state] = {}
        array_E = {}
        count_dic[state] = 0

# 对训练集获取状态标签
def get_tag(word):
    tag = []
    if len(word) == 1:
        tag = ['S']
    elif len(word) == 2:
        tag = ['B', 'E']
    else:
        num = len(word) - 2
        tag.append('B')
        tag.extend(['M'] * num)
        tag.append('E')
    return tag
# 将参数估计的概率取对数，对概率 0 取无穷小 -3.14e+100
def Prob_Array():
    for key in array_Pi:
        if array_Pi[key] == 0:
            array_Pi[key] = -3.14e+100
        else:
            array_Pi[key] = log(array_Pi[key] / line_num)
    for key0 in array_A:
        for key1 in array_A[key0]:
            if array_A[key0][key1] == 0.0:
                array_A[key0][key1] = -3.14e+100
            else:
                array_A[key0][key1] = log(array_A[key0][key1] / count_dic[key0])
```

```python
        # print(array_A)
        for key in array_B:
            for word in array_B[key]:
                if array_B[key][word] == 0.0:
                    array_B[key][word] = -3.14e+100
                else:
                    array_B[key][word] = log(array_B[key][word] /count_dic[key])

# 将字典转换成数组
def Dic_Array(array_b):
    tmp = np.empty((4,len(array_b['B'])))
    for i in range(4):
        for j in range(len(array_b['B'])):
            tmp[i][j] = array_b[STATES[i]][list(word_set)[j]]
    return tmp

# 判断一个字最大发射概率的状态
def dist_tag():
    array_E['B']['begin'] = 0
    array_E['M']['begin'] = -3.14e+100
    array_E['E']['begin'] = -3.14e+100
    array_E['S']['begin'] = -3.14e+100
    array_E['B']['end'] = -3.14e+100
    array_E['M']['end'] = -3.14e+100
    array_E['E']['end'] = 0
    array_E['S']['end'] = -3.14e+100

def dist_word(word0,word1,word2,array_b):
    if dist_tag(word0,array_b) == 'S':
        array_E['B'][word1] = 0
        array_E['M'][word1] = -3.14e+100
        array_E['E'][word1] = -3.14e+100
        array_E['S'][word1] = -3.14e+100
    return
# Viterbi算法求测试集最优状态序列
def Viterbi(sentence,array_pi,array_a,array_b):
    tab = [{}]   # 动态规划表
    path = {}

    if sentence[0] not in array_b['B']:
        for state in STATES:
            if state == 'S':
                array_b[state][sentence[0]] = 0
            else:
```

```python
                    array_b[state][sentence[0]] = -3.14e+100
    # 初始化 tab[0] 时刻的值
        for state in STATES:
            # print(array_pi[state])
            # print(array_b[state][sentence[0]])
            tab[0][state] = array_pi[state] + array_b[state][sentence[0]]
            # print(tab[0][state])
            # tab[t][state] 表示时刻 t 到达 state 状态的所有路径中，概率最大路径的概率值
            path[state] = [state]
        for i in range(1,len(sentence)):
            tab.append({})
            new_path = {}
            # if sentence[i] not in array_b['B']:
            # print(sentence[i-1],sentence[i])
            for state in STATES:
                if state == 'B':
                    array_b[state]['begin'] = 0
                else:
                    array_b[state]['begin'] = -3.14e+100
            for state in STATES:
                if state == 'E':
                    array_b[state]['end'] = 0
                else:
                    array_b[state]['end'] = -3.14e+100
            for state0 in STATES:
                items = []
                # if sentence[i] not in word_set:
                #     array_b[state0][sentence[i]] = -3.14e+100
                # if sentence[i] not in array_b[state0]:
                #     array_b[state0][sentence[i]] = -3.14e+100
                # print(sentence[i] + state0)
                # print(array_b[state0][sentence[i]])
                for state1 in STATES:
                    # if tab[i-1][state1] == -3.14e+100:
                    # continue
                    # else:
                    if sentence[i] not in array_b[state0]:    # 所有在测试集出现但没有
在训练集中出现的字符
                        if sentence[i-1] not in array_b[state0]:
                            prob = tab[i - 1][state1] + array_a[state1][state0] + array_b[state0]['end']
                        else:
                            prob = tab[i - 1][state1] + array_a[state1][state0] + array_b[state0]['begin']
```

```
                    # print(sentence[i])
                    # prob = tab[i-1][state1] + array_a[state1][state0] +
array_b[state0]['other']
                else:
                    prob = tab[i-1][state1] + array_a[state1][state0] +
array_b[state0][sentence[i]]            # 计算每个字符对应STATES的概率
    #                   print(prob)
                items.append((prob,state1))
            # print(sentence[i] + state0)
            # print(array_b[state0][sentence[i]])
            # print(sentence[i])
            # print(items)
            best = max(items)     #bset:(prob,state)
            # print(best)
            tab[i][state0] = best[0]
            # print(tab[i][state0])
            new_path[state0] = path[best[1]] + [state0]
        path = new_path

    prob, state = max([(tab[len(sentence) - 1][state], state) for state in STATES])
    return path[state]


# 根据状态序列进行分词
def tag_seg(sentence,tag):
    word_list = []
    start = -1
    started = False

    if len(tag) != len(sentence):
        return None

    if len(tag) == 1:
        word_list.append(sentence[0])       # 语句只有一个字，直接输出

    else:
        if tag[-1] == 'B' or tag[-1] == 'M': # 若最后一个字状态不是'S'或'E'则修改
            if tag[-2] == 'B' or tag[-2] == 'M':
                tag[-1] = 'S'
            else:
                tag[-1] = 'E'

        for i in range(len(tag)):
            if tag[i] == 'S':
                if started:
```

```python
                started = False
                word_list.append(sentence[start:i])
            word_list.append(sentence[i])
        elif tag[i] == 'B':
            if started:
                word_list.append(sentence[start:i])
            start = i
            started = True
        elif tag[i] == 'E':
            started = False
            word = sentence[start:i + 1]
            word_list.append(word)
        elif tag[i] == 'M':
            continue

    return word_list


if __name__ == '__main__':
    trainset = open('CTBtrainingset.txt', encoding='utf-8')      # 读取训练集
    testset = open('CTBtestingset.txt', encoding='utf-8')        # 读取测试集
    goldseg = open('CTB_test_gold.txt', encoding='utf-8')
    # trainlist = []

    Init_Array()# 初始化所有概率矩阵

    for line in trainset:
        line = line.strip()# 去除字符串首尾的字符
        # trainlist.append(line)
        line_num += 1 # 统计训练集语句的数量

        word_list = [] # 所有字的集合
        for k in range(len(line)):
            if line[k] == ' ':continue
            word_list.append(line[k])
        # print(word_list)
        word_set = word_set | set(word_list) #| 求并集    # 训练集所有字的集合

        line = line.split(' ')
        # print(line)
        line_state = []       # 这句话的状态序列

        for i in line:
            line_state.extend(get_tag(i))
```

```
        # print(line_state[0])
        array_Pi[line_state[0]] += 1          # array_Pi 用于计算初始状态分布概率

        for j in range(len(line_state)-1):
            # count_dic[line_state[j]] += 1       # 记录每一个状态的出现次数
            array_A[line_state[j]][line_state[j+1]] += 1   #array_A 计算状态转移概率

        for p in range(len(line_state)):
            count_dic[line_state[p]] += 1         # 记录每一个状态的出现次数
            for state in STATES:
                if word_list[p] not in array_B[state]:
                    array_B[state][word_list[p]] = 0.0   #保证每个字都在STATES的字典中

            # if word_list[p] not in array_B[line_state[p]]:
            #     # print(word_list[p])
            #     array_B[line_state[p]][word_list[p]] = 0
            # else:
            array_B[line_state[p]][word_list[p]] += 1   # array_B用于计算发射概率

Prob_Array()      # 对概率取对数保证精度

# print('参数估计结果')
# print('初始状态分布')
# print(array_Pi)
# print('状态转移矩阵')
# print(array_A)
# print('发射矩阵')
# print(array_B)
Precision = []
Recall = []
F1 = []
gold = next(goldseg)

output = ''

for line in testset:
    line = line.strip()
    gold = gold.strip()
    tag = Viterbi(line, array_Pi, array_A, array_B)
        # print(tag)

    seg = tag_seg(line, tag)


    list = ''
```

```
        for i in range(len(seg)):
            list = list + seg[i] + ' '
        dic = gold.split()
        output = output + list + '\n'
        print(gold)
        print(dic)
        print(list)
        x = prf(gold,list,dic)
        Precision.append(x[0])
        Recall.append(x[1])
        F1.append(x[2])
        # print("Precision:%.2f Recall:%.2f F1:%.2f " %prf(gold,list,dic))

        try:
            gold = next(goldseg)
        except StopIteration:
            break

print('准确率:')
print(mean(Precision))
print('召回率:')
print(mean(Recall))
print(mean(F1))

# print(output)
outputfile = open('output.txt', mode='w', encoding='utf-8')
outputfile.write(output)
```

代码运行结果：

外商 投资 企业 成为 中国 外贸 重要 增长 点好，我们需要休息一下
1997年 实现 进出口 总值 达 一千零九十八点二亿 美元

分词准确率结果如图4.6所示。

图4.6　分词准确率

实战应用：基于马尔可夫模型的文本生成器

应用背景：由于新媒体领域快速发展，公司业务量激增，结合业务需求，受"AI作画"启发，经理希望信息技术部能够开发一套"AI写作"信息系统，辅助媒体工作者进行新闻写作。待该系统成熟落地后，可以应用到学生写作教育，进一步扩展公司的业务范围。请结合所学知识，尽快开展

相关技术储备及落地应用。

1. 数据集

可以自己准备，越丰富越好，样例如下：

> 爱的雪花，染白了头发的四季，揉皱了面颊。与你一起互依互靠，为彼此的一切操心操劳。
> 爱过知情重，醉过知酒浓，缘分不停留，像春风来又走，徒留伤悲在心中，不知能与谁共。唯愿星斗作证，我心依旧！爱你，就舍不得看到你眼神的忧郁；爱你，就不忘记在你流泪的时候加倍安慰；爱你，就无怨无悔地把我的快乐一句一句放进你的心里，让你永远感到甜蜜。
> 你的笑温暖我的心，你的美灿烂一个季，你的快乐是我的心愿，你的幸福是我的春天，拥有你我春心荡漾，离开你我心从春出！你的快乐，我的甘甜；你的忧郁，我的悲伤。我虽不是你一切的一切，但愿做你永远的永远。七夕之日，知道我在想你吗？面貌的美丽固然重要，但心灵和思想的美丽才是真挚爱情的牢固基础。虽然你并不沉鱼落雁，但却有一颗金子般的心。每次想你在心头，彩虹也微笑开了口；每次想你在深夜，星星也为我守候。我们的爱，美了自然界，甜了心尖尖。满天星云下，感到无边的落寞。也许流星能体会。我期待，飘雪的日子，你的心带来了一缕芬芳。

> I have searched a thousand years, and I have cried a thousand tears. I found everything I need, You are everything to me.
> The revised Population and Family Planning Law, which allows Chinese couples to have three children, was passed by the Standing Committee of the National People's Congress, China's top legislature, on Friday.
> The amended law stipulates that the country will take supportive measures, including those in finances, taxes, insurance, education, housing arld employment, to reduce families' burdens as well as the cost of raising and educating children.

2. 模型构建

```python
import jieba
import random
import re
from nltk import word_tokenize

import nltk
nltk.download('punkt')

class Markov:
    def __init__(self, language):
        # language: 'chinese' / english
        self.language = language.lower() # 语言
        self.file_path = None    # 训练文件
        
        self.chain = None # key-value字典序列
    
    # 从指定文件目录读取文本，给make_sentence使用，如 ['我爱你', '我爱玩']
    def read_file(self):
```

```python
        rule = "[_.!+-=——,$%^:,;。? 、~@#¥%……&*《》<>「」{}【】()/]"
        with open(self.file_path, 'r', encoding='utf-8') as f:
            data = f.readlines()
        for i in range(len(data)):
            data[i] = data[i].strip()
            data[i] = re.sub(rule, '', data[i])
        return data

    # 返回一个二维列表,比如 [['我','爱','你'],['我','爱','玩']]
    def make_sentence(self):
        # 从文件中读取原字符串
        str_lt = self.read_file()
        sentence_lt = []
        if self.language == 'chinese':
            for str in str_lt:
                sentence_lt.append(list(jieba.cut(str)))
        elif self.language == 'english':
            for str in str_lt:
                sentence = word_tokenize(str)
                sentence = [word.strip() for word in sentence]
                sentence = [word for word in sentence if len(word) > 0]
                sentence_lt.append(sentence)
        return sentence_lt

    # 使用前调用一下生成key-value的字典
    def learn(self, file_path):  # [[], []...]
        self.file_path = file_path   # 存储训练文件文件
        sentence_lt = self.make_sentence()
        chain = {}
        for sentence in sentence_lt:
            for index in range(1, len(sentence)):
                if sentence[index - 1] in chain:
                    chain[sentence[index - 1]].append(sentence[index])
                else:
                    chain[sentence[index - 1]] = [sentence[index]]
        self.chain = chain

    # 进行造句
    def create(self, length, firstwz_lt):
        if self.chain is None:
            print('error, please use train method to make the module learn the language')
            return None

        if firstwz_lt is None:
            start = random.choice(list(self.chain.keys()))
        else:
            start = random.choice(firstwz_lt)
        create_word_lt = [start]  # 组成的句子
        while len(create_word_lt) < length:
            start = random.choice(self.chain.get(start, list(self.chain.keys())))
            create_word_lt.append(start)
```

```
        # 根据不同语言合成句子
        ans = 'error'
        if self.language == 'chinese':
            ans = ''.join(create_word_lt)
        elif self.language == 'english':
            ans = ' '.join(create_word_lt)
        return ans
```

3. 模型测试

```
markov = Markov('english')

markov.learn('/content/raw_en.txt')
# 造一个以 I 开头的 6 个单词的句子
markov.create(6, ['I'])
I have cried a thousand years

markov = Markov('chinese')

markov.learn('/content/raw_ch.txt')
# 造一个以 " 我 " 开头的 6 个词的句子
markov.create(6, [' 我 '])
我在想你就无怨无悔
```

单元小结

了解隐马尔可夫的基础原理以及应用，对于了解自然语言处理问题的基本思想和技术发展脉络有很大的好处。从本单元的学习可以看出，解决序列标注及序列生成类问题时隐马尔可夫模型表现出了优异的性能。

本单元让同学们了解了中文分词的经典方法之———隐马尔可夫模型。首先大致介绍了隐马尔可夫链与二元语法的基本知识，然后通过实例让同学们理解到隐马尔可夫链是如何进行词性标注的，在此基础上详细说明了二元语法与隐马尔可夫链结合应用于中文分词之中的代码实现；之后介绍了隐马尔可夫模型的建立以及利用中文语料库对模型进行训练的过程；最后详细介绍了查重率与查准率，让同学们了解如何判断分词方法准确度并将其应用于对隐马尔可夫方法的准确度检测中。

下一单元将学习下一个经典模型——感知机。

习 题

1. 什么是隐马尔可夫链？
2. 什么是二元语法？

单元5 感知机

在上一单元中介绍了隐马尔可夫模型，该模型的学习算法是一种计数式的训练算法，通过统计训练集上各事件的发生次数，然后利用极大似然估计归一化频次后得到相关概率，这些概率就是学习到的模型参数。而本单元要学习的感知机算法是一种迭代式的算法，在训练集上运行多次迭代，每次读入一个样本，执行预测，将预测结果与正确答案进行对比，计算误差，根据误差更新模型参数。迭代的次数一般是人工指定的一个参数，这个参数称作超参数。

感知机模型虽然比较简单，但为之后的工作提供了很好的思路，是支持向量机和神经网络的基础，并且在二分类和结构化预测问题上都有不错的表现。本单元知识导图如图5.1所示。

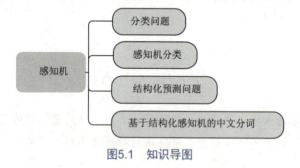

图5.1 知识导图

课程安排

课程任务	课程目标	安排学时
掌握分类问题	了解分类问题，并能正确区分分类问题	1
掌握感知机分类方式	了解感知机的定义及原理，并能完成简单代码应用	1
了解结构化预测问题	了解结构化预测问题的定义并能够区分分类与预测问题	1
掌握感知机在中文分词中的应用	掌握如何应用感知机解决中文分词问题，并能完成简单代码应用	2
实战应用	通过该实战应用来引导同学们了解一些业务背景，解决实际问题，进行自我练习、自我提高	2

5.1 分类问题

分类问题（Classification）是人工智能领域的一个重要组成部分，在机器学习中，最常见的问题就是分类问题。所谓分类问题，就是根据已知样本的某些特征，判断一个新的样本属于哪种已知的样本类。根据样本类的数量，可以将分类问题进一步划分为二分类、三分类、四分类等，习惯上将其称为二分类（Binary Classification）和多分类问题（Multiclass Classification）。

例如，根据一个人的体温来判断这个人是否发热，这就是一个典型的二分类问题（发热或者未发热）；电子邮箱在收到邮件之后，会将邮件分为广告邮件、垃圾邮件和正常邮件，这就是一个多分类问题，如图5.2所示。

图5.2 分类问题

5.2 基于分类的感知机分类

对于二分类问题，可以通过感知机来解决。感知机（Perceptron）是一种线性分类模型，主要解决二分类问题。它通过将输入空间的数据线性划分（正负两类）来完成分类。感知机的学习方式是基于分类错误的点，通过误分类的点来构建损失函数，利用梯度下降原理对损失函数进行调优，即极小化损失，从而得到线性划分的超平面。

感知机基本拥有神经网络的主要构件与思想。其模型公式为

$$f(x) = \text{sign}(w \cdot x + b)$$

接下来我们举一个二维平面下的二分类的例子来理解感知机模型。

如图 5.3 所示，左侧区域和右侧区域的点分别表示两种类别，需要将两种颜色的点划分开来。下面看一下感知机是如何做到的。

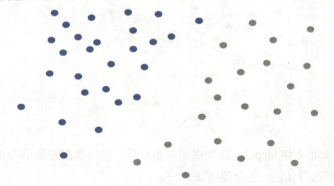

图5.3 两种颜色的点分类

Step 1：首先根据模型公式生成一条直线，初始直线具有随机性，可以是图 5.4 中任意一种情况。

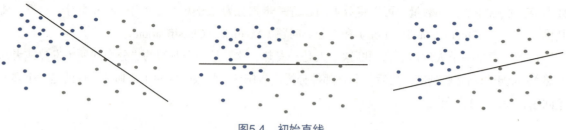

图5.4 初始直线

Step 2：假定图 5.4 中最右边的情况是本次生成的初始直线，根据此直线划分会有多个误分点。

如图 5.5 所示，圈中的部分表示此次划分中被分类错误的点，此时有 10 个分类错误的点。接下来，感知机根据预测错误的点来对划分直线进行调整、优化。

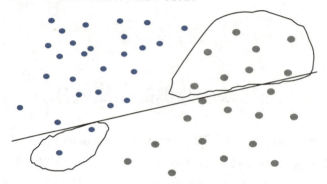

图5.5 划分误差

Step 3：根据分类错误的点对划分直线进行调整。

如图 5.6 所示，虚直线表示上一次划分的结果，实线表示优化后的结果。在这一轮调整后，分类错误的点只有 5 个，如图 5.7 所示。

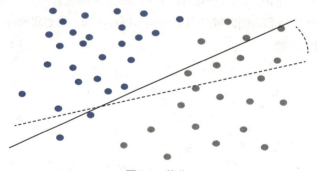

图5.6 优化

此后不断地循环 Step 2 和 Step 3，直到找到一条直线，使分类误差最小为止。最终结果如图 5.8 所示，找到一条直线将两种颜色的点全部划分。

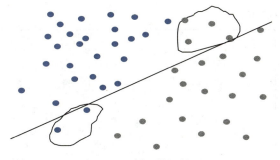

图5.7 新一轮误差

图5.8 最终结果

代码实践：

首先导入需要的库。

```
from sklearn.linear_model import Perceptron
import numpy as np
from sklearn.datasets import make_classification    # 分类数据集
import matplotlib.pyplot as plt
```

其中 make_classification 为 sklearn 中生成分类数据集的方法，生成二分类数据集代码如下：

```
# 定义数据集
X, y = make_classification(
    n_samples=100,              # 样本数量
    n_features=2,               # 特征总量
    n_informative=2,            # 有效特征数量
    n_redundant=0,              # 无效特征
    n_classes=2,                # 类别
    n_clusters_per_class=1,     # 每一个类别为一个簇
    random_state=8
)
# 画浮点图
plt.scatter(X[:, 0], X[:, 1], c=y)
plt.show()
```

生成的浮点图如图 5.9 所示。

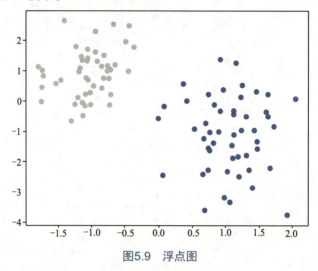

图5.9　浮点图

实例化感知机模型并拟合数据后，观察生成的切分直线。

```
# 实例化感知机
clf = Perceptron()
clf.fit(X, Y)
# 画出感知机的切分线
plt.scatter(X[:, 0], X[:, 1], c=y)
x_ = np.arange(-2, 2)
y_ = -(clf.coef_[0][0]*x_ + clf.intercept_) / clf.coef_[0][1]
plt.plot(x_, y_)
plt.show()
```

生成的切分直线如图 5.10 所示。

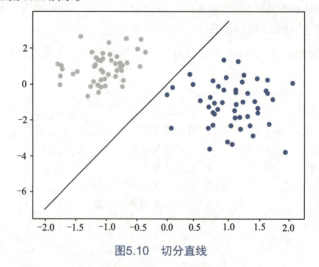

图5.10　切分直线

以上就是感知机在分类问题上的基本思路以及在 Python 中的用法。感知机于 1957 年由 Rosenblatt 提出。其被证实存在一定的缺陷。

例如，对于图 5.11，无法找一个直线来对两种颜色的点进行有效划分，所以说感知机无法解决线性不可分的问题。

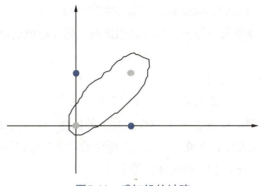

图5.11　感知机的缺陷

5.3　结构化预测问题

机器学习中会遇到各种各样的问题，在诸多问题当中，分类问题和回归问题是最常见的。

分类问题在 5.1 节已经介绍过了，回归问题就是要通过一组特征预测一个具体的值，一个标量。例如，通过父母的身高，来预测孩子的身高，这就是一种典型的回归问题。

除了上述两种问题之外，可能还需要输出一个具有结构化的结果，如一个序列、一棵树、一个列表，此时就需要用到结构化学习。例如，在中英文机器翻译任务中，需要输入一段英文，然后输出一段与之对应的中文，即输入一段语言序列，输出另一段语言序列。又如，在语音识别任务中，输入一段音频，会输出识别的语言序列。微信聊天中就会应用到语音识别功能，我们说一段话，对应一段音频信号，可以将所说的话转化为一段识别出的文字，也就对应一段语言序列。以上都属于结构化预测问题。

结构化学习可以解决结构化预测问题，也就是说输出不仅有值，还有与之对应的结构。结构化学习有一个通用的框架，具体如下：

Step 1：设计一个打分函数 score(x,y)，其中 x 为输入，y 为输出。打分函数可以计算输入和输出之间的分数，分数越高说明 y 越准确。

Step 2：输入 x 后，从所有可能的 y 中找到一个 y'，使 y' 和 x 之间的分数最大，然后把这个 y' 当作最后的预测结果，即

$$y = \arg\max \text{sore}(x, y'), y' \in Y$$

5.4　基于结构化感知机的中文分词

在 2.1 节已经介绍了分词。那么中文分词和感知机有什么联系呢？连接中文分词与感知机的桥梁，就是基于字标注的分词方法。

分词由切分问题转化为序列标注问题。每个字都有一个标注，即每个字属于一个类别，那么给

出一个字的标注过程其实就是一个分类过程。所以，利用感知机可以解决这样的问题。

以常用的 4-tag 标注系统为例，假如规定每个字最多有四个构词位置，即 B（词首）、M（词中）、E（词尾）和 S（单独成词）。这就是 4-tag 标注系统中的四个位置标注。

那么，对于任意一个已经分词的句子，都可以用这四个标注组成的序列，表示原来的分词结果。例如：

分词结果：我 / 爱 / 北京 / 天安门 / 。/

字标注形式：我 /S 爱 /S 北 /B 京 /E 天 /B 安 /M 门 /E 。/S

需要注意的是，这里的字标注形式，并不仅仅是对字的标注，也包括文本中出现的任何一个字符。因为在真实的文本中，会出现大量的字符，所以这里所说的"字"也包括字符。

接下来看一下实例：基于结构化感知机的中文分词。

```python
import zipfile
import os

from pyhanlp import *
from pyhanlp.static import download, remove_file, HANLP_DATA_PATH


def test_data_path():
    """
    获取测试数据路径，位于 $root/data/test，根目录由配置文件指定。
    :return:
    """
    data_path = os.path.join(HANLP_DATA_PATH, 'test')
    if not os.path.isdir(data_path):
        os.mkdir(data_path)
    return data_path


# 验证是否存在 MSR 语料库，如果没有则自动下载
def ensure_data(data_name, data_url):
    root_path = test_data_path()
    dest_path = os.path.join(root_path, data_name)
    if os.path.exists(dest_path):
        return dest_path

    if data_url.endswith('.zip'):
        dest_path += '.zip'
    download(data_url, dest_path)
    if data_url.endswith('.zip'):
        with zipfile.ZipFile(dest_path, "r") as archive:
            archive.extractall(root_path)
        remove_file(dest_path)
```

```python
        dest_path = dest_path[:-len('.zip')]
    return dest_path


sighan05 = ensure_data('icwb2-data', 'http://sighan.cs.uchicago.edu/bakeoff2005/data/icwb2-data.zip')
msr_train = os.path.join(sighan05, 'training', 'msr_training.utf8')
msr_model = os.path.join(test_data_path(), 'msr_cws')
msr_test = os.path.join(sighan05, 'testing', 'msr_test.utf8')
msr_output = os.path.join(sighan05, 'testing', 'msr_bigram_output.txt')
msr_gold = os.path.join(sighan05, 'gold', 'msr_test_gold.utf8')
msr_dict = os.path.join(sighan05, 'gold', 'msr_training_words.utf8')

# ================================================
# 以下开始中文分词

CWSTrainer = JClass('com.hankcs.hanlp.model.perceptron.CWSTrainer')
CWSEvaluator = SafeJClass('com.hankcs.hanlp.seg.common.CWSEvaluator')
HanLP.Config.ShowTermNature = False    # 关闭显示词性


def train_uncompressed_model():
    model = CWSTrainer().train(msr_train, msr_train, msr_model, 0., 10, 8).getModel()  # 训练模型
    model.save(msr_train, model.featureMap.entrySet(), 0, True)  # 最后一个参数指定导出 txt


def train():
    model = CWSTrainer().train(msr_train, msr_model).getModel()  # 训练感知机模型
    segment = PerceptronLexicalAnalyzer(model).enableCustomDictionary(False)  # 创建感知机分词器
    print(CWSEvaluator.evaluate(segment, msr_test, msr_output, msr_gold, msr_dict))  # 标准化评测
    return segment


segment = train()
sents = [
    "王××, 男, 1949年10月生。",
    "山东桓台县起凤镇穆寨村妇女穆××",
    "现为中国艺术研究院中国文化研究所研究员。",
    "北京输气管道工程",
]
for sent in sents:
    print(segment.seg(sent))
```

分词结果如图 5.12 所示。

```
P:96.68 R:96.51 F1:96.59 OOV-R:71.54 IV-R:97.18
[王××, , , 男, , , 1949年10月, 生, 。]
[山东, 桓台县, 起凤镇, 穆寨村, 妇女, 穆××]
[现, 为, 中国艺术研究院中国文化研究所, 研究员, 。]
[北京输气管道, 工程]
```

图5.12 分词结果

实战应用：使用感知机根据人名实现性别分类

应用背景：你所在公司的人力资源部由于工作疏忽，在进行员工信息登记时部分员工的性别信息缺失，为了降低工作成本，公司希望信息技术部通过 AI 手段利用人名实现性别预测。

请结合所学知识，尽快开展相关技术储备及落地应用。

数据集：使用 HanLP 下载中文人名性别数据集 cnname。其中包含 20 万条训练集，以及 2 万条测试集，分别标注了姓名以及性别。数据样例见表 5.1。

表5.1 数据格式

序号	姓名	性别	序号	姓名	性别
1	田有东	男	6	孙爱仙	女
2	缪剑钢	男	7	杨莲菊	女
3	李浩垣	男	8	赵东祥	男
4	张杜超	男	9	邢丽炫	女
5	于光浦	男	10	张为梅	女

首先需要导入要使用的工具包。HanLP 是一系列模型与算法组成的 NLP 工具包，其中封装了大量的 API 对应各种任务。只需要调用接口即可。

```python
from pyhanlp import *
from pyhanlp.static import download, remove_file, HANLP_DATA_PATH
import zipfile
import os
```

之后需要导入数据集。以下是定义的导入数据集函数，其中 test_data_path() 函数用来返回存储 cnname 数据集的位置。也就是在当前工程文件下新建一个 test 文件夹来保存 cnname 数据集。

```python
def test_data_path():
    """
    获取测试数据路径，位于 $root/data/test，根目录由配置文件指定
    """
    data_path = os.path.join(HANLP_DATA_PATH, 'test')
    if not os.path.isdir(data_path):
        os.mkdir(data_path)
```

```
        return data_path

def ensure_data(data_name, data_url):
"""# 验证是否存在cnname人名性别语料库，如果没有则自动下载"""
    root_path = test_data_path()
    dest_path = os.path.join(root_path, data_name)
    if os.path.exists(dest_path):
        return dest_path

    if data_url.endswith('.zip'):
        dest_path += '.zip'
    download(data_url, dest_path)
    if data_url.endswith('.zip'):
        with zipfile.ZipFile(dest_path, "r") as archive:
            archive.extractall(root_path)
        remove_file(dest_path)
        dest_path = dest_path[:-len('.zip')]
    return dest_path
```

ensure_data() 这个函数是用来验证是否已经下载该数据集，如果没有该数据集则下载并解压。以下代码是以上两个函数的调用以及感知机人名性别分类模型的加载。

```
# 读取人名性别分类模型
PerceptronNameGenderClassifier = JClass('com.hankcs.hanlp.model.perceptron.PerceptronNameGenderClassifier')
cnname = ensure_data('cnname', 'http://file.hankcs.com/corpus/cnname.zip')
TRAINING_SET = os.path.join(cnname, 'train.csv')
TESTING_SET = os.path.join(cnname, 'test.csv')
MODEL = cnname + ".bin"
```

该方法执行后生成目录如图 5.13 所示，其中 train.csv 是训练集，test.csv 是测试数据集。

名称	修改日期	类型
LICENSE	2022/8/9 17:17	文件
README.md	2022/8/9 17:17	MD 文件
README	2022/8/9 17:17	Internet 快捷方式
test.csv	2022/8/9 17:17	Microsoft Excel 逗...
train.csv	2022/8/9 17:17	Microsoft Excel 逗...

图5.13　目录结构

以下是模型的实例化以及准确度的测试。此处比较了两种感知机算法，分别是平均感知机算法和朴素感知机算法。

```
def run_classifier(averaged_perceptron):
    print('=====%s=====' % ('平均感知机算法' if averaged_perceptron else '朴素感知机算法'))
```

```
# 实例化感知机模型
classifier = PerceptronNameGenderClassifier()
print('训练集准确率:', classifier.train(TRAINING_SET, 10, averaged_perceptron))
model = classifier.getModel()
print('特征数量:', len(model.parameter))
# model.save(MODEL, model.featureMap.entrySet(), 0, True)
# classifier = PerceptronNameGenderClassifier(MODEL)
for name in "王冰冰", "李建军", "张鹏", "李世龙":
    print('%s=%s' % (name, classifier.predict(name)))
print('测试集准确率:', classifier.evaluate(TESTING_SET))
```

在主函数中调用上述方法。run_classifier() 方法中填入 False 表示使用朴素感知机算法,True 表示平均感知机算法。

```
if __name__ == '__main__':
    run_classifier(False)
    run_classifier(True)
```

运行结果如图 5.14 所示。

```
=====朴素感知机算法=====
训练集准确率:   P=84.56 R=87.24 F1=85.88
特征数量:   9089
王冰冰=女
李建军=男
张鹏=男
李世龙=男
测试集准确率:   P=82.09 R=85.01 F1=83.52
=====平均感知机算法=====
训练集准确率:   P=93.56 R=83.04 F1=87.98
特征数量:   9089
王冰冰=女
李建军=男
张鹏=男
李世龙=男
测试集准确率:   P=90.90 R=80.48 F1=85.37
```

图5.14　运行结果

以下是完整代码:

```
from pyhanlp import *
from pyhanlp.static import download, remove_file, HANLP_DATA_PATH


import zipfile
import os




def test_data_path():
    """
```

```
        获取测试数据路径,位于$root/data/test,根目录由配置文件指定
        """
        data_path = os.path.join(HANLP_DATA_PATH, 'test')
        if not os.path.isdir(data_path):
            os.mkdir(data_path)
        return data_path


    def ensure_data(data_name, data_url):
        """# 验证是否存在cnname人名性别语料库,如果没有则自动下载"""
        root_path = test_data_path()
        dest_path = os.path.join(root_path, data_name)
        if os.path.exists(dest_path):
            return dest_path

        if data_url.endswith('.zip'):
            dest_path += '.zip'
        download(data_url, dest_path)
        if data_url.endswith('.zip'):
            with zipfile.ZipFile(dest_path, "r") as archive:
                archive.extractall(root_path)
            remove_file(dest_path)
            dest_path = dest_path[:-len('.zip')]
        return dest_path


    # 读取人名性别分类模型
    PerceptronNameGenderClassifier = JClass('com.hankcs.hanlp.model.perceptron.PerceptronNameGenderClassifier')
    cnname = ensure_data('cnname', 'http://file.hankcs.com/corpus/cnname.zip')
    TRAINING_SET = os.path.join(cnname, 'train.csv')
    TESTING_SET = os.path.join(cnname, 'test.csv')
    MODEL = cnname + ".bin"


    def run_classifier(averaged_perceptron):
        print('=====%s=====' % ('平均感知机算法' if averaged_perceptron else '朴素感知机算法'))

        # 实例化感知机模型
        classifier = PerceptronNameGenderClassifier()
        print('训练集准确率:', classifier.train(TRAINING_SET, 10, averaged_perceptron))
        model = classifier.getModel()
```

```
        print('特征数量：', len(model.parameter))
    # model.save(MODEL, model.featureMap.entrySet(), 0, True)
    # classifier = PerceptronNameGenderClassifier(MODEL)
    for name in "王冰冰", "李建军", "张鹏", "李世龙":
        print('%s=%s' % (name, classifier.predict(name)))
    print('测试集准确率：', classifier.evaluate(TESTING_SET))


if __name__ == '__main__':
    run_classifier(False)
    run_classifier(True)
```

单 元 小 结

自然语言处理的问题大致可分为两类：一种是分类问题，另一种是结构化预测问题，而感知机既能解决分类问题又能处理结构化预测问题。对感知机稍作拓展，分类器就能支持结构化预测。

本单元介绍了分类问题以及感知机结构化预测问题，并且分别叙述了分类问题的常见场景和结构化预测问题的常见场景。学习了感知机模型，并且通过代码实践的形式让同学们有了更深刻的理解，其中介绍了 sklearn 工具包的使用，包括 linear_model 和 datasets 的用法。常见的机器学习模型大都可以从 sklearn 中调用，只不过有时需要对模型调参。最后做了一个基于感知机的人名性别分类，又重温了之前介绍的 HanLP 工具的使用。同学们需要动手实践，以加深理解。

下一单元将学习功能更为强大的基础模型——条件随机场模型。

习 题

1．什么是感知机？
2．如何用感知机实现性别分类应用？

单元6 条件随机场

本单元介绍一种新的自然语言处理模型——条件随机场，这种模型与感知机同属结构化学习大家族，但性能比感知机还要强大。条件随机场常用于序列标注、词性标注等自然语言处理任务中。

本单元将重点介绍条件随机场的应用，若想探究条件随机场的原理，需要先了解马尔可夫模型以及概率图模型。此外，本单元还将介绍条件随机场工具——CRF++，以及CRF++在NLP领域的使用。本单元知识导图如图6.1所示。

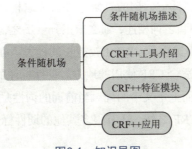

图6.1 知识导图

课程安排

课程任务	课程目标	安排学时
了解条件随机场	了解条件随机场的定义及应用范围	1
掌握CRF++工具	能够灵活运用CRF++工具，并能解决实际问题	1
实战应用	通过该实战应用来引导同学们了解一些业务背景，解决实际问题，进行自我练习、自我提高	2

6.1 条件随机场描述

条件随机场（Conditional Random Field，CRF）是一种机器学习模型，在2001年由Lafferty等提出，结合了最大熵模型和隐马尔可夫模型的特点，属于随机场的一种，在NLP领域常用于分词、词性标注、命名实体识别。

条件随机场是条件概率分布模型 $P(Y|X)$，表示的是给定一组输入随机变量 X 的条件下另一组输出随机变量 Y 的马尔可夫随机场。例如，在词性标注问题上，输入一个句子，那么输出就是这个句子并且在每个词之后标注了词性。下面通过一个CRF常见的例子来了解它的用途。

假设有小王一天从早到晚的一系列照片，如工作的照片，吃饭的照片，唱歌的照片等，需要根据这些照片判断小王的具体活动。一个比较直观的办法就是，找到小王之前日常生活的一系列照片，然后找小王问清楚这些照片代表的活动标记，这样就可以用监督学习方法来训练一个分类模型，接着用模型去预测这一天的每张照片最可能的活动标记。这种办法虽然可行，却忽略了一个重要的问题，就是这些照片之间的顺序。在自然语言处理领域，输入的文本大多是有时序性的，一般的网络结构无法提取时序性信息，这也是为什么循环神经网络（RNN）及其各种变体在自然语言处理领域比较流行的原因。

继续回到刚才小王的问题，比如现在看到了一张小王闭着嘴的照片，那么这张照片应怎么标记小王的活动呢？单单一张闭嘴照片很难去打标记。但是如果有小王在这一张照片前一点点时间的照片，那么这张照片就好标记了。如果在时间序列上前一张的照片里小王在吃饭，那么这张闭嘴的照片很有可能是在咀嚼；而如果在时间序列上前一张的照片里小王在唱歌，那么这张闭嘴的照片很有可能也是在唱歌。

为了让分类器表现得更好，在标记数据的时候，可以考虑相邻数据的标记信息。具体来说，可以设置一个打分函数，假如说动词和名词相连，那么给正分。反之，如果两个动词相连，那么给负分。这是普通的分类器难以做到的。而这也是 CRF 比较擅长的地方。在实际应用中，自然语言处理中的词性标注 (POS Tagging) 就是非常适合 CRF 使用的地方。词性标注的目标是给出一个句子中每个词的词性（名词、动词、形容词等）。而上下文中的词对当前词的词性是有相关性的，因此，使用 CRF 处理会有很好的效果。当然，除了 CRF 之外，还有很多其他的词性标注方法。

6.2 CRF++工具

CRF++ 是一个可用于分词、连续数据标注的简单、可定制的开源 CRF 工具，也是目前综合性能最佳的 CRF 工具。CRF++ 是为了通用目的而设计的，能被用于自然语言信息处理的各个方面，诸如命名实体识别、信息提取和语块分析等。

CRF++ 的安装非常简单，在 Windows 环境下，只需要从官网下载 CRF++ 的安装压缩文件，然后解压到某个目录下面即可，图 6.2 为解压后的目录。

下面对其中的文件 / 文件夹进行介绍：

（1）doc 文件夹：官方主页的内容。

（2）example 文件夹：有四个数据包，每个数据包有训练数据（test.data）、测试数据 (train.data)、模板文件 (template)、执行脚本文件（exec.sh）四个文件。

（3）sdk 文件夹：CRF++ 的头文件和静态链接库。

（4）clr_learn.exe：CRF++ 的训练程序。

（5）crl_test.exe：CRF++ 的测试程序。

（6）libcrffpp.dll：训练程序和测试程序需要使用的静态链接库。

（7）其他文件为 CRF++ 的一些说明信息，如 CRF++ 官方网站，它的著作人以及版权信息等。

名称	修改日期	类型
doc	2013/2/12 23:40	文件夹
example	2013/2/12 23:40	文件夹
sdk	2013/2/12 23:40	文件夹
AUTHORS	2013/2/12 23:40	文件
BSD	2013/2/12 23:40	文件
COPYING	2013/2/12 23:40	文件
crf_learn.exe	2013/2/12 23:40	应用程序
crf_test.exe	2013/2/12 23:40	应用程序
LGPL	2013/2/12 23:40	文件
libcrfpp.dll	2013/2/12 23:40	应用程序扩展
README	2013/2/12 23:40	文件

图6.2　Windows下解压后的CRF++目录

6.3　CRF++特征模板

训练和测试文件必须包含多个 token（每一行就代表一个 token），每个 token 又包含多个列。每个 token 必须写在一行，且各列之间用空格或制表格间隔。一个 token 的序列可构成一个句子，每个句子之间用一个空行间隔。

下面看一下官方文档给出的样例。

例句：He reckons the current account deficit will narrow to only #1.8 billion in September。

图 6.3 是这句话的格式示例。

```
He         PRP    B-NP
reckons    VBZ    B-VP
the        DT     B-NP
current    JJ     I-NP
account    NN     I-NP
deficit    NN     I-NP
will       MD     B-VP
narrow     VB     I-VP
to         TO     B-PP
only       RB     B-NP
#          #      I-NP
1.8        CD     I-NP
billion    CD     I-NP
in         IN     B-PP
September  NNP    B-NP
.          .      O

He         PRP    B-NP
reckons    VBZ    B-VP
..
```

图6.3　格式示例

"He reckons the …"代表了一个训练语句,CRF++会将该语句拆开,每一行有一个单词,并且固定列数。除了包含原始信息,可能还包含了其他信息,图6.3中的第二列就表示单词词性信息。最后一列是标记信息,也就是label。不同的句子通过空白行来分隔。第一个token中,第一列He表示原始单词,第二列PRP表示单词词性——代词,具体的词性信息可以查看标注词表,最后一列就是序列标注的信息,B-NP表示的是名字短语的开头,可以参考BIO标注。

为了更加清晰地解释上例,把使用到的标注词以及BIO标注进行说明如下:

名词:NN,NNS,NNP,NNPS

代词:PRP,PRP$

形容词:JJ,JJR,JJS

数词:CD

动词:VB,VBD,VBG,VBN,VBP,VBZ

副词:RB,RBR,RBS

BIO标注:

B-X:属于X类型的开头,B-NP表名词开头,B-VP表动词开头。

I-X:属于X类型的中间,I-BP表示名词中间,B-VP表动词中间。

O:不属于任何类型。

模板就是规则,通过一个模板来约束相应的条件。由于CRF++被设计为通用工具,因此必须提前指定特征模板。特征模板是用来配置特征的。模板文件中的每一行表示一个模板。在每个模板中,将使用特殊的%x(row,col)在输入数据中指定标记,row指定当前焦点标记的相对位置,col指定列的绝对位置。

继续看官方文档中的例子,假定在这句话中当前词是the,那么表6.1就是使用模板的表示方法。

表6.1 数据结构

Template	Expanded feature	Template	Expanded feature
%x[0,0]	the	%x[-2,1]	PRP
%x[0,1]	DT	%x[0,0]/%x[0,1]	the/DT
%x[-1,0]	reckons	ABC%x[0,1]123	ABCDT123

下面详细解读表6.1中的例子。

%x[0,0]:因为当前词是the,而且[0,0]中的第一个0表示当前位置(行),第二个0表示第一列,所以%x[0,0]依旧表示the这个单词。

%x[0,1]:表示当前词的行数和第二列,对应表6.1中的DT。

%x[-1,0]:表示当前词的前一行和第一列的位置,即reckons。

ABC%x[0,1]123:表示字符'ABC'拼接上%x[0,1]位置上的单词,再拼接上'123',因为%x[0,1]代表DT,所以这里表示ABCDT123。

CRF++模板构建分为两类,一类是Unigram标注,一类是Bigram标注。

Unigram template: 首字符: 'U'

这是一个描述单一语法特征的模板。当给出模板"U01:%x[0，1]"时，CRF++会自动生成一组功能函数（func1 ...funcN）：

```
func1=if(output=B-NP and feature="U01:DT")return 1 else return 0
func2=if(output=I-NP and feature="U01:DT")return 1 else return 0)
func3=if(output=0 and feature="U01:DT")return 1 else return 0))
...
funcXX=if(output=B-NP and feature="U01:NN")return 1 else return 0
funcXY=if(output=0 and feature="U01:NN")return 1 else return 0
...
```

这些函数反映了训练样例的情况，func1反映了训练样例中词性是DT且标注是B-NP的情况，func2反映了训练样例中词性是DT且标注是I-NP的情况，依此类推。

```
Bigram template: 首字符: 'B'
```

这是一个描述双层函数的模板。使用此模板，将自动生成当前输出令牌和以前的输出令牌（双帧）的组合。

6.4 CRF++命令行预测

可以使用CRF++自带的例子进行一个测试。例如，在example中chunking文件夹下进行一个测试，其中有四个文件：exec.sh、template、test.data和train.data。将crf_learn.exe、crf_test.exe和libcrfpp.dll三个文件复制到这个文件夹（chunking）下，如图6.4所示。

名称	修改日期	类型
crf_learn.exe	2013/2/12 23:40	应用程序
crf_test.exe	2013/2/12 23:40	应用程序
exec.sh	2013/2/12 23:40	Shell Script
libcrfpp.dll	2013/2/12 23:40	应用程序扩展
template	2013/2/12 23:40	文件
test.data	2013/2/12 23:40	DATA 文件
train.data	2013/2/12 23:40	DATA 文件

图6.4 复制之后example中的chunking目录

之后在命令行执行命令进行训练。首先切换到chunking文件夹下，如图6.5所示，其中D:\System Soft\CRF++-0.58\example\chunking是此时计算机中chunking的路径，同学们需要将此路径改为自己计算机中对应的路。

```
Microsoft Windows [版本 10.0.22000.856]
(c) Microsoft Corporation。保留所有权利。

C:\Users\zb>D:

D:\>cd D:\System Soft\CRF++-0.58\example\chunking

D:\System Soft\CRF++-0.58\example\chunking>
```

图6.5 切换chunking路径

切换到 chunking 路径之后，执行以下代码来训练模型。在命令窗口中，执行 crf_learn template train.data model 命令，具体如图 6.6 所示。

```
D:\System Soft\CRF++-0.58\example\chunking>crf_learn template train.data model
CRF++: Yet Another CRF Tool Kit
Copyright (C) 2005-2013 Taku Kudo, All rights reserved.

reading training data:
Done!0.02 s

Number of sentences: 77
Number of features:  153482
Number of thread(s): 16
Freq:                1
eta:                 0.00010
C:                   1.00000
shrinking size:      20
iter=0  terr=0.98629 serr=1.00000 act=153482 obj=5003.65270 diff=1.00000
iter=1  terr=0.38449 serr=1.00000 act=153482 obj=4082.64911 diff=0.18407
iter=2  terr=0.38502 serr=1.00000 act=153482 obj=1974.35839 diff=0.51640
iter=3  terr=0.19937 serr=0.93506 act=153482 obj=1323.56081 diff=0.32962
iter=4  terr=0.14030 serr=0.92208 act=153482 obj=826.22783 diff=0.37575
iter=5  terr=0.08333 serr=0.71429 act=153482 obj=573.51229 diff=0.30587
iter=6  terr=0.03692 serr=0.44156 act=153482 obj=388.80427 diff=0.32206
iter=7  terr=0.01266 serr=0.20779 act=153482 obj=310.94172 diff=0.20026
iter=8  terr=0.00158 serr=0.03896 act=153482 obj=285.42807 diff=0.08205
iter=9  terr=0.00105 serr=0.02597 act=153482 obj=273.98088 diff=0.04011
iter=10 terr=0.00000 serr=0.00000 act=153482 obj=266.16068 diff=0.02854
iter=11 terr=0.00000 serr=0.00000 act=153482 obj=260.00330 diff=0.02313
iter=12 terr=0.00000 serr=0.00000 act=153482 obj=257.78354 diff=0.00854
iter=13 terr=0.00000 serr=0.00000 act=153482 obj=256.04494 diff=0.00674
iter=14 terr=0.00000 serr=0.00000 act=153482 obj=255.43711 diff=0.00237
iter=15 terr=0.00000 serr=0.00000 act=153482 obj=254.99790 diff=0.00172
iter=16 terr=0.00000 serr=0.00000 act=153482 obj=254.37422 diff=0.00245
iter=17 terr=0.00000 serr=0.00000 act=153482 obj=254.69603 diff=0.00127
iter=18 terr=0.00000 serr=0.00000 act=153482 obj=254.17182 diff=0.00206
iter=19 terr=0.00000 serr=0.00000 act=153482 obj=254.02268 diff=0.00059
iter=20 terr=0.00000 serr=0.00000 act=153482 obj=253.92179 diff=0.00040
iter=21 terr=0.00000 serr=0.00000 act=153482 obj=253.85845 diff=0.00025
iter=22 terr=0.00000 serr=0.00000 act=153482 obj=253.85970 diff=0.00000
iter=23 terr=0.00000 serr=0.00000 act=153482 obj=253.82884 diff=0.00012
iter=24 terr=0.00000 serr=0.00000 act=153482 obj=253.80341 diff=0.00010
iter=25 terr=0.00000 serr=0.00000 act=153482 obj=253.79206 diff=0.00004
iter=26 terr=0.00000 serr=0.00000 act=153482 obj=253.78214 diff=0.00004
iter=27 terr=0.00000 serr=0.00000 act=153482 obj=253.77759 diff=0.00002

Done!0.55 s
```

图6.6 执行训练

这里还有四个参数可以调整，在之后的实践部分会有使用。

1. -a CRF-L2 or CRF-L1

规范化算法选择。默认是 CRF-L2。一般来说 L2 算法效果要比 L1 算法稍好，虽然 L1 算法中非零特征的数值要比 L2 中的小。

2. -c float

这个参数设置 CRF 的 hyper-parameter。c 的数值越大，CRF 拟合训练数据的程度越高。这个参数可以调整过度拟合和不拟合之间的平衡度。这个参数可以通过交叉验证等方法寻找较优的参数。

3. -f NUM

这个参数设置特征的 cut-off threshold。CRF++ 使用训练数据中至少 NUM 次出现的特征。默认值为 1。当使用 CRF++ 到大规模数据时，只出现一次的特征可能会有几百万，这个选项就会在这样的情况下起到作用。

4. -p NUM

如果计算机中有多个 CPU，那么可以通过多线程提升训练速度。NUM 是线程数量。

可以看到控制台打印了很多信息，并且在 chunking 文件下产生了一个新的文件 model。

输出参数的含义如下：

- iter：处理的迭代次数。
- terr：与标签有关的错误率（错误标记的个数/所有标记的个数）。
- serr：句子的错误率（错误句子数/所有句子数）。
- obj：当前对象值。当此值收敛到固定点时，CRF++ 将停止迭代。
- diff：与上一个对象值的相对差异。

可以看出，随着训练迭代次数的增加，错误率会逐渐下降，也就使模型越来越准确。下面结合实例介绍 CRF++ 工具使用。

训练数据集：backof2005

Step 1: 准备好数据集，使用 /icwb2-data/training/msr_training.utf8 作为训练数据集。由于在 windows 下 CRF++ 并不支持大数据运算，所以这里使用了该数据集一半的数据量，并且需要将数据集转化为 CRF++ 训练所需要的格式。采用 4-tag（B，E，M，S）标记。

B：Begin，词首

E：End，词尾

M：Middle，词中

S：Single，单子词

以下片段为 msr_training.utf8 数据中的部分数据。

```
"  人们  常  说  生活  是  一  部  教科书  ，  而  不  平凡  的  经历
   是  其中  不可多得  的  部分  ，  她  确实  是  名副其实  的  '  我  的  大学  '  。
"  心静渐知春似海  ，  花深每觉影生香  。
```

以下代码将文本调整为 CRF++ 训练格式。

```
import codecs
```

```python
def character_tagging(input_file, output_file):
    input_data = codecs.open(input_file, 'r', 'utf-8')
    output_data = codecs.open(output_file, 'w', 'utf-8')
    for line in input_data.readlines():
        word_list = line.strip().split()
        for word in word_list:
            if len(word) == 1:
                output_data.write(word + "\tS\n")
            else:
                output_data.write(word[0] + "\tB\n")
                for w in word[1:len(word)-1]:
                    output_data.write(w + "\tM\n")
                output_data.write(word[len(word)-1] + "\tE\n")
        output_data.write("\n")
    input_data.close()
    output_data.close()

if __name__ == '__main__':
    character_tagging(r" msr_training.utf8", "msr_training.tagging4crf.utf8")
```

加入了 4-tag（B，E，M，S）标记，并且调整为 token 形式。注意这里 character_tagging() 方法中传入的参数，第一个参数为 msr_training.utf8 数据的位置，第二个参数为生成的数据的名字，以下为调整之后的格式，如图 6.7 所示。

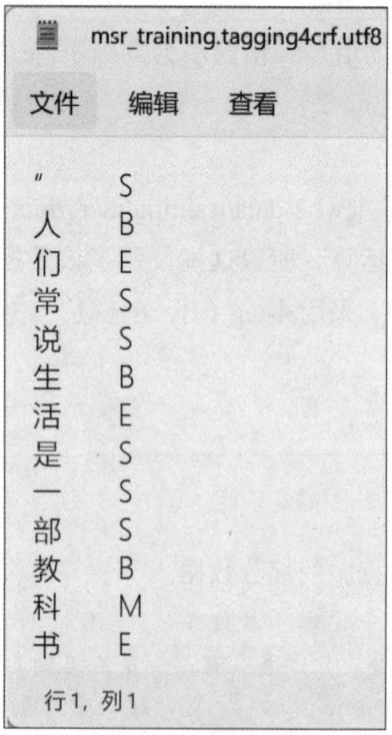

图6.7　数据格式

Step 2：转化为 CRF++ 的训练格式之后就可以训练模型了。可以新建一个文件夹来保存数据和模型。这里新建了一个 my_demo 文件夹，并把 CRF++ 工具中的四个文件复制过来，如图 6.8 所示。

名称	修改日期	类型
crf_learn.exe	2013/2/12 23:40	应用程序
crf_test.exe	2013/2/12 23:40	应用程序
libcrfpp.dll	2013/2/12 23:40	应用程序扩展
template	2013/2/12 23:40	文件

图6.8　文件目录

需要注意的是，这里使用的 template 文件为 example，seg 文件下的 template。将上述代码生成的 msr_training.tagging4crf.utf8 数据文件也复制到该文件夹下。有了工具，有了数据，就可以通过命令行来训练模型。

通过命令行跳转到 my_demo 文件下后输入命令 crf_learn -f 3 -c 4.0 template msr_training.tagging4crf.utf8 model（这里设置了两个参数）。命令执行后会开始读数据并训练，运行结果如图 6.9 所示。

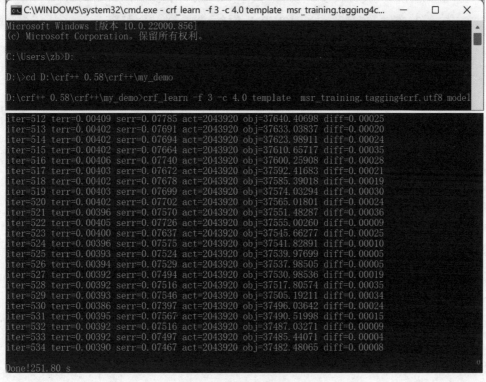

图6.9　运行结果

执行完成后会生成我们需要的模型文件 model，如图 6.10 所示。

名称	修改日期	类型
crf_learn.exe	2013/2/12 23:40	应用程序
crf_test.exe	2013/2/12 23:40	应用程序
libcrfpp.dll	2013/2/12 23:40	应用程序扩展
model	2022/8/15 18:46	文件
msr_training.tagging4crf.utf8	2022/8/15 18:42	UTF8 文件
template	2013/2/12 23:40	文件

图6.10　存储后的模型

Step 3：有了模型之后就可以分词了，但是在这之前，需要将测试数据集也改为 CRF++ 需要的格式。

```python
import codecs

def character_split(input_file, output_file):
    input_data = codecs.open(input_file, 'r', 'utf-8')
    output_data = codecs.open(output_file, 'w', 'utf-8')
    for line in input_data.readlines():
        for word in line.strip():
            word = word.strip()
            if word:
                output_data.write(word + "\tB\n")
        output_data.write("\n")
    input_data.close()
    output_data.close()


if __name__ == '__main__':
    character_split(r" msr_test.utf8", "msr_test4crf.utf8")
```

使用 msr_test.utf8 作为测试集。这里 character_split() 的参数同上，转化后的部分测试集如图 6.11 所示。

Step 4：将测试集带入模型中，在命令行执行命令 crf_test -m model msr_test4crf.utf8 > msr_test4crf.tag.utf8，如图 6.12 所示。运行完成后会生成 msr_test4crf.tag.utf8 文件。打开该文件，发现已经得到标注的结果，如图 6.13 所示。

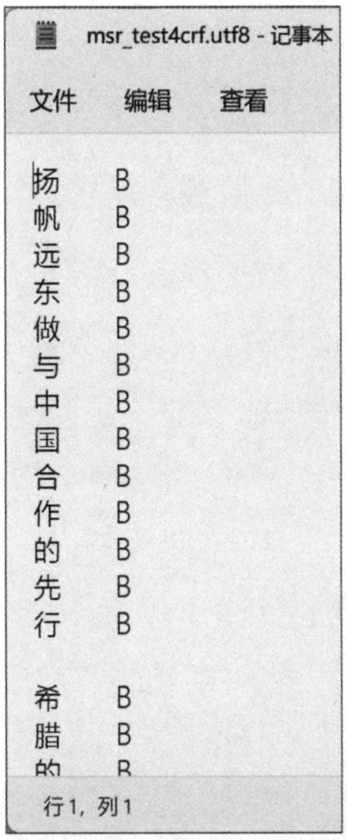

图6.11 数据结构

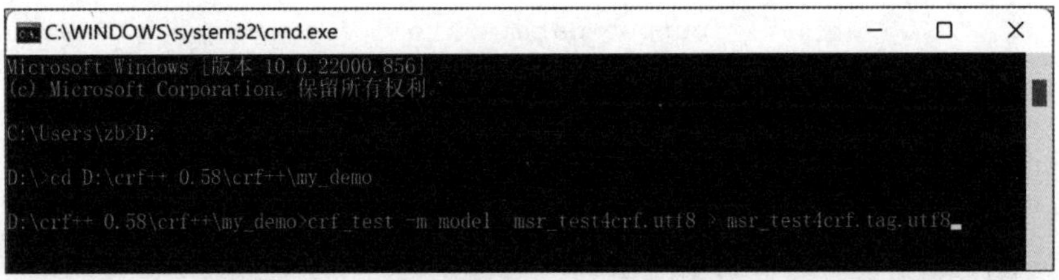

图6.12 运行结果

图6.13 结果示意图

Step 5：查看分词结果。

```
import codecs
```

```
def character_2_word(input_file, output_file):
    input_data = codecs.open(input_file, 'r', 'utf-8')
    output_data = codecs.open(output_file, 'w', 'utf-8')
    for line in input_data.readlines():

        if line.strip() == '':
            output_data.write('/n')
        else:
            char_tag_pair = line.strip().split('\t')
            char = char_tag_pair[0]
            tag = char_tag_pair[2]
            if tag == 'B':
                output_data.write(' ' + char)
            elif tag == 'M':
                output_data.write(char)
            elif tag == 'E':
                output_data.write(char + ' ')
            else:    # tag == 'S'
                output_data.write(' ' + char + ' ')
    input_data.close()
    output_data.close()
```

实战应用：基于条件随机场的词性标注

应用背景：还记得在单元4编写的"AI写作"软件吗？经过项目团队的不断迭代优化，目前已经成功落地并进入市场，产品上市后大受好评。为了进一步拓展软件的功能，经过客户经理调研发现客户对于文本的词性标注有较为迫切的需求。词性标注在本质上是分类问题，将语料库中的单词按词性分类。一个词的词性由其在所属语言的含义、形态和语法功能决定。以汉语为例，汉语的词类系统有18个子类，包括7类体词、4类谓词、5类虚词、代词和感叹词。请结合所学知识，尽快开展相关技术储备及落地应用。

1. 数据集

本案例使用PKU语料库，在程序执行过程中会自动下载。

2. 模型构建及训练

```
from pyhanlp import *
import zipfile
import os
from pyhanlp.static import download, remove_file, HANLP_DATA_PATH


def test_data_path():
    """
```

        获取测试数据路径, 位于$root/data/test, 根目录由配置文件指定。
        :return:
        """
        data_path = os.path.join(HANLP_DATA_PATH, 'test')
        if not os.path.isdir(data_path):
            os.mkdir(data_path)
        return data_path

    # 验证是否存在 MSR 语料库, 如果没有自动下载
    def ensure_data(data_name, data_url):
        root_path = test_data_path()
        dest_path = os.path.join(root_path, data_name)
        if os.path.exists(dest_path):
            return dest_path

        if data_url.endswith('.zip'):
            dest_path += '.zip'
        download(data_url, dest_path)
        if data_url.endswith('.zip'):
            with zipfile.ZipFile(dest_path, "r") as archive:
                archive.extractall(root_path)
            remove_file(dest_path)
            dest_path = dest_path[:-len('.zip')]
        return dest_path

    # 指定 PKU 语料库
    PKU98 = ensure_data("pku98", "http://file.hankcs.com/corpus/pku98.zip")
    PKU199801 = os.path.join(PKU98, '199801.txt')
    PKU199801_TRAIN = os.path.join(PKU98, '199801-train.txt')
    PKU199801_TEST = os.path.join(PKU98, '199801-test.txt')
    POS_MODEL = os.path.join(PKU98, 'pos.bin')
    NER_MODEL = os.path.join(PKU98, 'ner.bin')

    # 以下开始 CRF 词性标注

    AbstractLexicalAnalyzer = JClass('com.hankcs.hanlp.tokenizer.lexical.AbstractLexicalAnalyzer')
    PerceptronSegmenter = JClass('com.hankcs.hanlp.model.perceptron.PerceptronSegmenter')
    CRFPOSTagger = JClass('com.hankcs.hanlp.model.crf.CRFPOSTagger')
```
```

```python
def train_crf_pos(corpus):
    # 选项1. 使用 HanLP 的 Java API 训练，慢
    tagger = CRFPOSTagger(None)                    # 创建空白标注器
    tagger.train(corpus, POS_MODEL)                # 训练
    tagger = CRFPOSTagger(POS_MODEL)               # 加载
    analyzer = AbstractLexicalAnalyzer(PerceptronSegmenter(), tagger)  # 构造
词法分析器，与感知机分词器结合，能同时进行分词和词性标注。
    print(analyzer.analyze("张三的希望是希望上学"))    # 分词+词性标注
    print(analyzer.analyze("张三的希望是希望上学").translateLabels())  # 对词性
进行翻译
    return tagger

if __name__ == '__main__':
    tagger = train_crf_pos(PKU199801_TRAIN)
```

3. 模型测试

测试结果如图 6.14 所示。

```
张三/nr 的/u 希望/n 是/v 希望/v 上学/v
张三人名 的/助词 希望/名词 是/动词 希望/动词 上学/动词
```

图6.14 测试结果

单 元 小 结

条件随机场是自然语言处理中被广泛应用的一类模型，可简单描述为给定一组输入序列条件下另一组输出序列的条件概率分布模型。

本单元首先介绍了条件随机场的基本思想，又通过 CRF++ 工具训练了中文分词器。CRF++ 是非常强大的 CRF 工具，CRF++ 工具在 Python 中并没有库可以导入，需要从官网下载，并且在命令行运行。同学们也了解了分词的一些思想，了解了分词中的 4-tag（B，E，M，S）标记。值得注意的是，CRF++ 工具无法在 Windows 系统中大规模训练，自行实践时需要注意数据量。在实战应用中，主要学习如何运用条件随机场模型解决词性标注问题。

从下一单元开始，将学习自然语言处理中的典型子任务。

习 题

1. 什么是条件随机场？它在自然语言处理中有何用途？
2. 如何用条件随机场实现词性标注？

单元7 命名实体识别

从本单元开始进入自然语言处理子任务学习,下面将接触第一个子任务——命名实体识别。

命名实体识别是一种自然语言处理技术,可自动录入所提供的文本的全部内容,提取文本中的一些实体,并将它们进行分类。例如,晚上对手机语音助手说"定一个明天早上七点的闹钟",或者"下周三进行会议",然后手机语音助手将会根据你所提供的语句来进行相应的设置执行。手机语音助手是如何准确定位你所讲的话中的重点呢?这里用到的就是命名实体识别,它可以将用户所提供的文本进行分类,将其分成不同的实体,从而达到想要的效果。

本单元主要介绍命名实体识别的基础概念以及对命名实体识别运用时所需要的各种方式。要求同学们能基本掌握命名实体识别的算法原理及应用场景。本单元知识导图如图7.1所示。

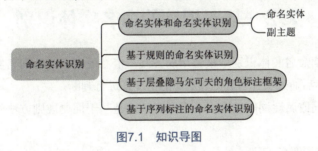

图7.1 知识导图

课程安排

课程任务	课程目标	安排学时
了解命名实体识别	了解命名实体和命名实体识别基本概念	1
了解基于规则的命名识别	了解基于规则的命名实体识别概念及其分类	1
了解角色标注框架	了解基于层叠马尔可夫模型的角色标注框架基本概念	1
掌握基于序列标注的命名实体识别	了解基于序列标注的命名实体识别原理,并能够运行代码解决简单应用	1
实战应用	通过该实战应用来引导同学们了解一些业务背景,解决实际问题,进行自我练习、自我提高	2

7.1 命名实体和命名实体识别

7.1.1 命名实体

狭义的命名实体是人名、机构名、地名以及其他所有以名称为标识的实体。而广义的命名实体还包括数字、日期、货币、地址等。

7.1.2 命名实体识别

命名实体识别是指在文档集合中识别出特定类型的事物名称或符号的过程。

命名实体识别要回答三个问题：如何识别出文本中的命名实体；怎样来确定该实体的类型；如何解决在对于多个实体表示同一事物时，正确选择出其中的一个实体作为该类型实体的代表。

命名实体识别多出现于自然语言处理任务之中，是自然语言处理技术由理论化转向实用化的过程中最重要的一个步骤。一般来说,命名实体识别的主要任务就是识别出待处理文本中三大类(实体类、时间类和数字类)和七小类（人名、机构名、地名、时间、日期、货币和百分比）的命名实体。

7.2 基于规则的命名实体识别

规则加词典是早期命名实体识别中最常用的方式。依靠规则，结合命名实体库，对每条规则进行不同的权重赋值，然后通过实体与规则相对应的情况来进行判断。

实体词表、关系词或属性词触发词表、正则表达式是基于词典规则方法的三大核心部件，主要分为两种识别方式：

1. 基于实体词表的匹配识别

基于实体词表的匹配识别是使用最广泛的一种实体识别方法，虽然实体词表实现目标文本词表的有限匹配，但见效十分快速。

在进行特定领域实体识别时，每个领域都有专属的实体词典，每一个专属实体词典都可以用来进行实体识别。而对于有歧义的词汇，可以先进行分词，再在分词的基础上进行命名实体识别任务。

2. 基于规则模板的匹配识别

规则模板可以实现对实体词表识别的扩展，这个识别方法的核心在于规则模板的设计，在进行这个方法之前需要对实体词或者属性值的构词规则进行分析，这个方法又分为两种：基于字符构词规则的识别以及基于词性组合规则的识别。

7.3 基于层叠隐马尔可夫模型的角色标注框架

角色标注框架是一个统计命名实体识别框架，进行命名实体的短词语标签化，使标签序列能够满足某种模式时则会被识别为某种命名实体。规则系统根据词典的匹配规则来确定，统计方法可根据隐马尔可夫模型的预测来确定。

该框架中角色标注模块的输入是分词模块的输出,两个模块都是由隐马尔可夫模型驱动,所以称为层叠隐马尔可夫模型。根据识别目标的不同,角色标注所使用的标注集也不同。

7.4 基于序列标注的命名实体识别

常用的标注方式为 BIO、BIOES、IOB 等。

BIO 标注:"B-X"表示该元素所处的部分属于 X 类型且此元素在该部分的开头,"I-X"表示该元素所处的部分属于 X 类型且该元素在此部分的中间位置,而"O"表示不属于任何类型。

BIOES 标注:B 表示开始,I 表示内部,O 表示非实体,E 表示实体尾部,S 表示该词为一个实体。

IOB 标注:IOB 与 BIO 字母相对应的含义相同,其中不同的地方在于 IOB 中,标签 B 仅用于两个连续的同样类型的命名实体的边界区分,不用于命名实体的起始位置。

相比之下,IOB 因缺少 B-X 作为实体标注的开头,故丢失了部分标注的信息,导致在很多任务上的所得到的标注效果并不理想,而 BIO 则较好地解决了 IOB 的这个问题,因此整体的标注效果要比 IOB 更加优秀。而 BIOES 则是提供了 End 的信息,并给出了单个词汇的 S,提供了更多的信息,因此在不同的任务里 BIO 和 BIOES 各有优劣,但总的来说两种方法的表现差异并不大。

下面结合实践介绍如何基于语料库序列标注解决命名实体识别问题。

数据集:使用人民日报 2014 语料库进行序列标注的命名实体识别。

实践环境:Python 3.9。

实践代码:

```python
import torch
import torch.nn as nn
import torch.optim as optim
from sklearn.metrics import accuracy_score


# 定义开始词和结束词
START_TAG = "<START>"
STOP_TAG = "<STOP>"
# embedding 的尺寸
EMBEDDING_DIM = 100
# hidden 的尺寸
HIDDEN_DIM = 100


# 设置随机种子
torch.manual_seed(1)


# 返回最大概率的值
def argmax(vec):
    _, idx = torch.max(vec, 1)
```

```python
        return idx.item()

# 打包
def prepare_sequence(seq, to_ix):
    idxs = [to_ix[w] for w in seq]
    return torch.tensor(idxs, dtype=torch.long)

# 计算最大分数值
def log_sum_exp(vec):
    max_score = vec[0, argmax(vec)]
    max_score_broadcast = max_score.view(1, -1).expand(1, vec.size()[1])
    return max_score + \
        torch.log(torch.sum(torch.exp(vec - max_score_broadcast)))

# BiLSTM_CRF模型定义
class BiLSTM_CRF(nn.Module):
    # 初始化部分
    def __init__(self, vocab_size, tag_to_ix, embedding_dim, hidden_dim):
        super(BiLSTM_CRF, self).__init__()
        self.embedding_dim = embedding_dim
        self.hidden_dim = hidden_dim
        self.vocab_size = vocab_size
        self.tag_to_ix = tag_to_ix
        self.tagset_size = len(tag_to_ix)

        self.word_embeds = nn.Embedding(vocab_size, embedding_dim)
        self.lstm = nn.LSTM(embedding_dim, hidden_dim // 2,
                            num_layers=1, bidirectional=True)

        # 将 LSTM 的输出映射到标签空间
        self.hidden2tag = nn.Linear(hidden_dim, self.tagset_size)

        # 转换参数矩阵。 条目 i,j 是转换 *to* i *from* j.
        self.transitions = nn.Parameter(
            torch.randn(self.tagset_size, self.tagset_size))

        # 这两句话执行了如下约束：永远不会转移到开始标签，也永远不会从停止标签开始转移
        self.transitions.data[tag_to_ix[START_TAG], :] = -10000
        self.transitions.data[:, tag_to_ix[STOP_TAG]] = -10000

        self.hidden = self.init_hidden()

    # 随机初始化隐藏层参数
```

```python
def init_hidden(self):
    return (torch.randn(2, 1, self.hidden_dim // 2),
            torch.randn(2, 1, self.hidden_dim // 2))

def _forward_alg(self, feats):
    # 用正演算法计算配分函数
    init_alphas = torch.full((1, self.tagset_size), -10000.)
    # START_TAG 包含所有的分数
    init_alphas[0][self.tag_to_ix[START_TAG]] = 0.

    # 在变量中换行
    forward_var = init_alphas

    # 遍历句子
    for feat in feats:
        # 这个时间步的前向张量
        alphas_t = []
        for next_tag in range(self.tagset_size):
            # 广播的发射分数：不分先后
            # 前面的标签
            emit_score = feat[next_tag].view(
                1, -1).expand(1, self.tagset_size)
            # trans_score 的第 i 个条目是转换到的分数
            # next_tag 来自 i
            trans_score = self.transitions[next_tag].view(1, -1)
            # next_tag_var 的第 i 项是在执行 log-sum-exp 之前的边缘 (i -> next_tag) 的值
            next_tag_var = forward_var + trans_score + emit_score
            # 这个标记的前向变量是所有分数的 log-sum-exp
            alphas_t.append(log_sum_exp(next_tag_var).view(1))
        forward_var = torch.cat(alphas_t).view(1, -1)
    terminal_var = forward_var + self.transitions[self.tag_to_ix[STOP_TAG]]
    alpha = log_sum_exp(terminal_var)
    return alpha

# 获得句子隐藏维度的特征
def _get_lstm_features(self, sentence):
    self.hidden = self.init_hidden()
    embeds = self.word_embeds(sentence).view(len(sentence), 1, -1)
    lstm_out, self.hidden = self.lstm(embeds, self.hidden)
    lstm_out = lstm_out.view(len(sentence), self.hidden_dim)
    lstm_feats = self.hidden2tag(lstm_out)
    return lstm_feats
```

```python
        # 计算当前句子的得分
    def _score_sentence(self, feats, tags):
        # 给出所提供的标记序列的得分
        score = torch.zeros(1)
        tags = torch.cat([torch.tensor([self.tag_to_ix[START_TAG]], dtype=torch.long), tags])
        for i, feat in enumerate(feats):
            score = score + \
                self.transitions[tags[i + 1], tags[i]] + feat[tags[i + 1]]
        score = score + self.transitions[self.tag_to_ix[STOP_TAG], tags[-1]]
        return score

    def _viterbi_decode(self, feats):
        backpointers = []

        # 在日志空间中初始化viterbi变量
        init_vvars = torch.full((1, self.tagset_size), -10000.)
        init_vvars[0][self.tag_to_ix[START_TAG]] = 0

        # 第i步的Forward_var保存第i-1步的viterbi变量
        forward_var = init_vvars
        for feat in feats:
            # 保持此步骤的后向指针
            bptrs_t = []
            # 保存此步骤的viterbi变量
            viterbivars_t = []

            for next_tag in range(self.tagset_size):

                # next_tag_var[i]保存标签i的viterbi变量
                next_tag_var = forward_var + self.transitions[next_tag]
                best_tag_id = argmax(next_tag_var)
                bptrs_t.append(best_tag_id)
                viterbivars_t.append(next_tag_var[0][best_tag_id].view(1))
            # 现在添加发射分数,并将forward_var赋值给集合的viterbi变量
            forward_var = (torch.cat(viterbivars_t) + feat).view(1, -1)
            backpointers.append(bptrs_t)

        # 过渡到STOP_TAG
        terminal_var = forward_var + self.transitions[self.tag_to_ix[STOP_TAG]]
        best_tag_id = argmax(terminal_var)
        path_score = terminal_var[0][best_tag_id]
```

```python
        # 跟随后向指针来解码最佳路径
        best_path = [best_tag_id]
        for bptrs_t in reversed(backpointers):
            best_tag_id = bptrs_t[best_tag_id]
            best_path.append(best_tag_id)
        # 去掉开始标记
        start = best_path.pop()
        # 完整性检查
        assert start == self.tag_to_ix[START_TAG]
        best_path.reverse()
        return path_score, best_path

    # 计算损失
    def neg_log_likelihood(self, sentence, tags):
        feats = self._get_lstm_features(sentence)
        forward_score = self._forward_alg(feats)
        gold_score = self._score_sentence(feats, tags)
        return forward_score - gold_score

    # 模型调用
    def forward(self, sentence):
        # 从 BiLSTM 获取排放分数
        lstm_feats = self._get_lstm_features(sentence)

        # 根据特征找到最佳路径
        score, tag_seq = self._viterbi_decode(lstm_feats)
        # 标签
        return tag_seq

if __name__ == "__main__":

    # 读取数据集
    with open("data/2014_corpus.txt", encoding="utf-8") as f:
        datas = f.readlines()

    # 存储特征和标签
    all_data = []
    all_label = []

    # 遍历数据
    for data in datas:
        for da in data.split(' ')[:-1]:
            # 去除异常值
```

```python
            if len(da.split('/')) != 2:
                continue
            all_data.append(da.split('/')[0])
            all_label.append(da.split('/')[1])

    # 取1/1000数据做训练,可以自由改变分母
    all_data = all_data[:int(len(all_data)/100)]
    all_label = all_label[:int(len(all_label) / 100)]

    # 划分训练集和测试集
    # 8:2
    train_size = int(0.8*len(all_data))
    # 训练集
    train_data = all_data[:train_size]
    train_label = all_label[:train_size]

    # 测试集
    test_data = all_data[train_size:]
    test_label = all_label[train_size:]

    # 设置batch_size
    batch_size = 512

    # 开始的下标
    start = 0

    # 打包训练集
    training_datas = []
    for i in range(start, len(train_data), batch_size):
        training_datas.append((train_data[start:start+batch_size], train_label[start:start+batch_size]))
        start += batch_size

    # 词编码
    word_to_ix = {}
    for word in all_data:
        if word not in word_to_ix:
            word_to_ix[word] = len(word_to_ix)

    # 标签编码
    tag_to_ix = {}
    i = 0
    for tag in list(set(all_label)):
```

```python
        tag_to_ix[tag] = i
        i += 1
tag_to_ix[START_TAG] = i
tag_to_ix[STOP_TAG] = i+1

# 转换测试集标签
test_label = [tag_to_ix[label] for label in test_label]

# 初始化模型
model = BiLSTM_CRF(len(word_to_ix), tag_to_ix, EMBEDDING_DIM, HIDDEN_DIM)

# 定义优化器
optimizer = optim.SGD(model.parameters(), lr=0.01, weight_decay=1e-4)

print('开始模型的训练')
# 确保加载LSTM部分前面的prepare_sequence
# 训练10个epoch
for epoch in range(10):
    for sentence, tags in training_datas:
        # 清除梯度
        model.zero_grad()

        # 把它们变成单词索引的张量
        sentence_in = prepare_sequence(sentence, word_to_ix)

        # 把它们变成标签索引的张量
        targets = torch.tensor([tag_to_ix[t] for t in tags], dtype=torch.long)

        # 计算损失函数
        loss = model.neg_log_likelihood(sentence_in, targets)

        # 更新梯度
        loss.backward()

        optimizer.step()

    # 测试集上验证
    with torch.no_grad():
        # 把测试集数据变成单词索引的张量
        precheck_sent = prepare_sequence(test_data, word_to_ix)
        pred_result = model(precheck_sent)
        print(f'模型训练第{epoch+1}轮的准确率：{accuracy_score(pred_result, test_label)}')
```

模型训练结果如图 7.2 所示。

```
开始模型的训练
模型训练第1轮的准确率: 0.7137861766237631
模型训练第2轮的准确率: 0.768990402499814
模型训练第3轮的准确率: 0.7945093371028942
模型训练第4轮的准确率: 0.8097983780968678
模型训练第5轮的准确率: 0.8177590953054088
模型训练第6轮的准确率: 0.8255710140614537
模型训练第7轮的准确率: 0.8299605684100886
模型训练第8轮的准确率: 0.832266944423778
模型训练第9轮的准确率: 0.8336433301093669
模型训练第10轮的准确率: 0.8350197157949557
```

图7.2 模型训练结果

实战应用：热点问题命名实体识别

应用背景：你所在的新媒体公司每天需要处理海量的新闻数据，并需要从这些新闻数据中挖掘热点新闻。为了解决这一问题，首先需要识别出问题发生的地点、涉及的任务以及问题本身，支撑后续对热点问题的挖掘。公司将这一任务交给了你所在的团队，请根据所学知识解决相关问题。

1. 数据集

数据集是微软亚洲研究院提供的词性标注数据集，其目标是识别文本中具有特定意义的实体，包括人名、地名、机构名。

数据集示意图如图 7.3 所示。

```
有 的 长 篇 小 说 ， 经 过 惨 淡 经 营 而 斐 然 成 章 ， 自 饶 情 致 ， 独 具 魅 力 ， 或 因 哲 理 之 透 辟 ， 或 因 思 致 之 奇 巧 ， 或 因 文 字 之 粹 美 ， 或 因 故 事 之 曲 折 ， 以 摄 人 心 魄 ， 引 人 “ 凝 视 ” 。
我 为 影 片 的 主 人 公 徐 振 海 设 计 了 一 位 八 十 多 岁 的 老 母 亲 ， 儿 子 在 外 奔 波 挣 扎 一 天 ， 晚 上 回 到 家 还 记 着 给 老 娘 用 热 水 泡 泡 脚 。
黎 明 村 地 处 旱 塬 ， 人 均 耕 地 不 足 1 亩 。
```

图7.3 数据集示意图

2. 数据处理

（1）读取数据。

```
import codecs

# 训练数据和标签
train_lines = codecs.open('msra/train/sentences.txt').readlines()
train_lines = [x.replace(' ', '').strip() for x in train_lines]
# 用于移除字符串开头和结尾指定的字符（默认为空格或换行符）或字符序列

train_tags = codecs.open('msra/train/tags.txt').readlines()
train_tags = [x.strip().split(' ') for x in train_tags]
```

```
    train_tags = [[tag_type.index(x) for x in tag] for tag in train_tags]
    train_lines, train_tags = train_lines[:20000], train_tags[:20000]  # 只取两万数据
    print(train_lines[0], "\n", train_tags[0])

    # 验证数据和标签
    val_lines = codecs.open('msra/val/sentences.txt').readlines()
    val_lines = [x.replace(' ', '').strip() for x in val_lines]
    val_tags = codecs.open('msra/val/tags.txt').readlines()
    val_tags = [x.strip().split(' ') for x in val_tags]
    val_tags = [[tag_type.index(x) for x in tag] for tag in val_tags]
```

（2）数据分词。

```
    tokenizer = BertTokenizer.from_pretrained('bert-base-chinese')
    # 中文注意加 list(train_lines)

    max_length = 64
    train_encoding = tokenizer.batch_encode_plus(list(train_lines), truncation=True,
padding=True, max_length=max_length)
    val_encoding = tokenizer.batch_encode_plus(list(val_lines), truncation=True,
padding=True, max_length=max_length)
```

（3）定义数据读取。

```
    class TextDataset(Dataset):
        def __init__(self, encodings, labels):
            self.encodings = encodings
            self.labels = labels

        def __getitem__(self, idx):
            item = {key: torch.tensor(value[idx][:maxlen]) for key, value in
self.encodings.items()}
            # 字级别的标注
            item['labels'] = torch.tensor([0] + self.labels[idx] + [0] * (maxlen
- 1 - len(self.labels[idx])))[:maxlen]
            return item

        def __len__(self):
            return len(self.labels)


    train_dataset = TextDataset(train_encoding, train_tags)
    test_dataset = TextDataset(val_encoding, val_tags)
    print(train_dataset[0])

    # Dataset 转换成 Dataloader
```

```
batchsz = 32
train_loader = DataLoader(train_dataset, batch_size=batchsz, shuffle=True)
test_loader = DataLoader(test_dataset, batch_size=batchsz, shuffle=True)
```

3. 模型构建

```
from transformers import BertForTokenClassification, AdamW, get_linear_schedule_with_warmup

# 定义模型
model = BertForTokenClassification.from_pretrained('bert-base-chinese', num_labels=7)
device = torch.device("cuda:2" if torch.cuda.is_available() else "cpu")
model.to(device)

# 优化器和学习率
optimizer = AdamW(model.parameters(), lr=5e-5)
total_steps = len(train_loader) * 1
scheduler = get_linear_schedule_with_warmup(optimizer, num_warmup_steps=0,
                                            num_training_steps=total_steps)

# 这里对计算准确率进行测试
a = torch.tensor([1, 2, 3, 4, 2])
b = torch.tensor([1, 2, 4, 3, 2])
print((a==b).float().mean())
print((a==b).float().mean().item())
```

4. 模型训练

```
from tqdm import tqdm

def train():
    model.train()
    total_train_loss = 0
    iter_num = 0
    total_iter = len(train_loader)
    for idx, batch in enumerate(train_loader):
        optim.zero_grad()
        
        input_ids = batch['input_ids'].to(device)
        attention_mask = batch['attention_mask'].to(device)
        labels = batch['labels'].to(device)
        outputs = model(input_ids, attention_mask=attention_mask, labels=labels)
        # loss = outputs[0]
```

```python
            loss = outputs.loss

            if idx % 20 == 0:
                with torch.no_grad():
                    # 64 * 7
                    print((outputs[1].argmax(2).data == labels.data).float().mean().item(), loss.item())

            total_train_loss += loss.item()
            loss.backward()
            torch.nn.utils.clip_grad_norm_(model.parameters(), 1.0)
            optim.step()
            scheduler.step()

            iter_num += 1
            if(iter_num % 100==0):
                print("epoth: %d, iter_num: %d, loss: %.4f, %.2f%%" % (epoch, iter_num, loss.item(), iter_num/total_iter*100))

        print("Epoch: %d, Average training loss: %.4f"%(epoch, total_train_loss/len(train_loader)))

    def validation():
        model.eval()
        total_eval_accuracy = 0
        total_eval_loss = 0
        for batch in test_dataloader:
            with torch.no_grad():
                input_ids = batch['input_ids'].to(device)
                attention_mask = batch['attention_mask'].to(device)
                labels = batch['labels'].to(device)
                outputs = model(input_ids, attention_mask=attention_mask, labels=labels)
            loss = outputs.loss
            logits = outputs[1]

            total_eval_loss += loss.item()
            logits = logits.detach().cpu().numpy()
            label_ids = labels.to('cpu').numpy()
            total_eval_accuracy += (outputs[1].argmax(2).data == labels.data).float().mean().item()

        avg_val_accuracy = total_eval_accuracy / len(test_dataloader)
        print("Accuracy: %.4f" % (avg_val_accuracy))
```

```
        print("Average testing loss: %.4f"%(total_eval_loss/len(test_dataloader)))
        print("--------------------------------")


for epoch in range(4):
    print("------------Epoch: %d ----------------" % epoch)
    train()
    validation()
```

5. 模型预测

```
    model = torch.load('bert-ner.pt')

    tag_type = ['O', 'B-ORG', 'I-ORG', 'B-PER', 'I-PER', 'B-LOC', 'I-LOC']

    def predcit(s):
        item = tokenizer([s], truncation=True, padding='longest', max_length=64)
# 加一个list
        with torch.no_grad():
            input_ids = torch.tensor(item['input_ids']).to(device).reshape(1, -1)
            attention_mask = torch.tensor(item['attention_mask']).to(device).
reshape(1, -1)
            labels = torch.tensor([0] * attention_mask.shape[1]).to(device).
reshape(1, -1)

            outputs = model(input_ids, attention_mask, labels)
            outputs = outputs[0].data.cpu().numpy()

        outputs = outputs[0].argmax(1)[1:-1]
        ner_result = ''
        ner_flag = ''

        for o, c in zip(outputs,s):
            # 0 就是 O, 没有含义
            if o == 0 and ner_result == '':
                continue

            #
            elif o == 0 and ner_result != '':
                if ner_flag == 'O':
                    print('机构: ', ner_result)
                if ner_flag == 'P':
                    print('人名: ', ner_result)
                if ner_flag == 'L':
```

```
                print('位置: ', ner_result)
                ner_result = ''
        elif o != 0:
            ner_flag = tag_type[o][2]
            ner_result += c
    return outputs

s = '明天我们一起在海淀吃个饭吧,把叫刘涛和王华也叫上。'
data = predcit(s)
```

作为扩展学习,本单元还提供 BiLSTM 的 pytorch 实现版本,感兴趣的同学可以运行测试。

```python
# !/usr/bin/env python
# -*- encoding: utf-8 -*-
import os
import tqdm
import torch
import torch.nn as nn
from torch.utils.data import Dataset, DataLoader
from sklearn.metrics import f1_score


def build_corpus(split, make_vocab=True, data_dir="data"):
    """ 读取数据 """
    assert split in ["train", "dev", "test"]
    word_lists, tag_lists = [], []
    with open(os.path.join(data_dir, split + ".char.bmes"), mode="r", encoding="utf-8") as f:
        word_list, tag_list = [], []
        for line in f:
            if line != "\n":
                word, tag = line.strip("\n").split()
                word_list.append(word)
                tag_list.append(tag)
            else:
                word_lists.append(word_list)
                tag_lists.append(tag_list)
                word_list, tag_list = [], []
    word_lists = sorted(word_lists, key=lambda x: len(x), reverse=False)
    tag_lists = sorted(tag_lists, key=lambda x: len(x), reverse=False)

    # 如果 make_vocab 为 True,还需要返回 word2id 和 tag2id
    if make_vocab:
```

```python
            word2id = build_map(word_lists)
            tag2id = build_map(tag_lists)
            word2id['<UNK>'] = len(word2id)
            word2id['<PAD>'] = len(word2id)
            tag2id['<PAD>'] = len(tag2id)
            return word_lists, tag_lists, word2id, tag2id
        else:
            return word_lists, tag_lists


def build_map(lists):
    maps = {}
    for list_ in lists:
        for e in list_:
            if e not in maps:
                maps[e] = len(maps)
    return maps


class MyDataset(nn.Module):
    """ 自定义 Dataset 类 """

    def __init__(self, datas, tags, word2index, tag2index):
        self.datas = datas
        self.tags = tags
        self.word2index = word2index
        self.tag2index = tag2index

    def __getitem__(self, index):
        data = self.datas[index]
        tag = self.tags[index]

        data_index = [self.word2index.get(i, self.word2index['<UNK>']) for i in data]
        tag_index = [self.tag2index[i] for i in tag]

        return data_index, tag_index

    def __len__(self):
        assert len(self.datas) == len(self.tags)
        return len(self.tags)

    def pro_batch_data(self, batch_datas):
        """ 每个 batch 如何自动填充 """
```

```python
        global device
        datas, tags, batch_lens = [], [], []
        for data, tag in batch_datas:
            datas.append(data)
            tags.append(tag)
            batch_lens.append(len(data))
        batch_max_len = max(batch_lens)
        datas = [i + [self.word2index['<PAD>']] * (batch_max_len - len(i)) for i in datas]
        tags = [i + [self.tag2index['<PAD>']] * (batch_max_len - len(i)) for i in tags]

        return torch.tensor(datas, dtype=torch.int64, device=device), torch.tensor(tags, dtype=torch.long,device=device)   # long也是int64


class MyModel(nn.Module):
    def __init__(self, corpus_num, embedding_num, hidden_num, class_num, bi=True):
        super().__init__()
        self.embedding = nn.Embedding(corpus_num, embedding_num)
        self.lstm = nn.LSTM(embedding_num, hidden_num, batch_first=True, bidirectional=bi)

        if bi:
            self.classifer = nn.Linear(hidden_num * 2, class_num)
        else:
            self.classifer = nn.Linear(hidden_num, class_num)
        self.cross_loss = nn.CrossEntropyLoss()

    def forward(self, batch_data, batch_tag=None):
        embedding = self.embedding(batch_data)
        out, _ = self.lstm(embedding)
        pred = self.classifer(out)
        self.pred = torch.argmax(pred, dim=-1).reshape(-1)
        if batch_tag is not None:
            loss = self.cross_loss(pred.reshape(-1, pred.shape[-1]), batch_tag.reshape(-1))
            return loss


def test():
    global word2index, model, index2tag, device   # 全局变量声明，只是读取
    while True:
```

```python
        text = input("请输入：")
        text_index = [[word2index.get(i, word2index['<UNK>']) for i in text]]
        text_index = torch.tensor(text_index, dtype=torch.int64, device=device)
        model.forward(text_index)
        pred = [index2tag[i] for i in model.pred]

        print([f'{w}_{s}' for w, s in zip(text, pred)])


if __name__ == '__main__':
    device = "cuda:0" if torch.cuda.is_available() else "cpu"

    train_word_lists, train_tag_lists, word2index, tag2index = build_corpus("train", make_vocab=True)
    dev_data, dev_tag = build_corpus("dev", make_vocab=False)
    index2tag = [i for i in tag2index]

    # 定义变量
    corpus_num = len(word2index)
    class_num = len(tag2index)    # 命名实体识别就是为每个字进行分类
    epoch = 50
    lr = 0.001
    embedding = 101
    hidden_num = 107
    bi = True
    batchsz = 64

    train_dataset = MyDataset(train_word_lists, train_tag_lists, word2index, tag2index)
    # 自己处理：collate_fn=train_dataset.pro_batch_data
    train_dataloader = DataLoader(train_dataset, batch_size=batchsz, shuffle=False,
                                  collate_fn=train_dataset.pro_batch_data)

    dev_dataset = MyDataset(dev_data, dev_tag, word2index, tag2index)
    dev_dataloader = DataLoader(dev_dataset, batch_size=batchsz, shuffle=False,
                                collate_fn=dev_dataset.pro_batch_data)

    model = MyModel(corpus_num, embedding, hidden_num, class_num, bi)
    opt = torch.optim.Adam(model.parameters(), lr=lr)
    model = model.to(device)

    for e in tqdm.trange(epoch):
        model.train()
        for batch_data, batch_tag in train_dataloader:
```

```
            train_loss = model(batch_data, batch_tag)
            train_loss.backward()
            opt.step()
            opt.zero_grad()
        print(f"train loss: {train_loss:.3f}")

        model.eval()
        all_pred, all_tag = [], []
        for dev_batch_data, dev_batch_tag in dev_dataloader:
            dev_loss = model(dev_batch_data, dev_batch_tag)
            all_pred.extend(model.pred.detach().cpu().numpy().tolist())
            all_tag.extend(dev_batch_tag.detach().cpu().numpy().reshape(-1).tolist())
        # print(f"dev loss: {dev_loss:.3f}")
        score = f1_score(all_tag, all_pred, average="macro")
        print(f"{e},f1_score:{score:.3f},dev_loss:{dev_loss:.3f}")
    # test()
```

单 元 小 结

命名实体识别广泛应用于信息检索领域,在自然语言处理中占据着非常重要的地位。命名实体识别的学术理论和研究方法众多,本单元侧重于整体介绍。

本单元主要阐述了命名实体识别的基础知识和命名实体识别的常用方法,并在实战中运用命名实体识别方法支撑热点问题挖掘。

下一单元将在本单元的基础上学习信息抽取任务。

习 题

1. 什么是命名实体识别?它的应用范围是什么?
2. 命名实体识别的常用方法有哪些?它们的优缺点是什么?

单元8 信息抽取

上一单元介绍了如何解决命名实体识别任务,在此基础上,本单元将介绍如何解决信息抽取任务。

信息抽取,即从自然语言文字中,提取特定事件或事实信息,帮助完成自动分类、提取和重构海量内容的自然语言处理方法。信息抽取目前已经应用于很多领域,如商业智能、简历收获、媒体分析、情感检测、专利检索及电子邮件扫描。当前研究的一个特别重要的领域是提取出新闻的结构化数据,特别是在数字媒体行业。

本单元要求同学们了解信息抽取的各种方法,并能够将其运用到实际应用中。本单元知识导图如图 8.1 所示。

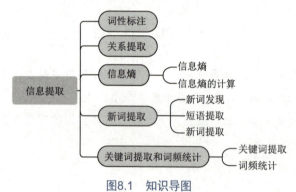

图8.1 知识导图

课程安排

课程任务	课程目标	安排学时
了解词性标注	了解词性标注问题的基本概念	1
了解关系提取	了解关系提取的概念及方法	1
掌握信息熵	了解信息熵的概念,掌握信息熵的计算方法	1
掌握新词提取	了解新词提取的概念,掌握短语提取和新词提取方法	1
掌握关键词提取和词频统计	掌握关键词提取和词频统计方法,并能运用代码解决简单问题	1
实战应用	通过该实战应用来引导同学们了解一些业务背景,解决实际问题,进行自我练习、自我提高	2

8.1 词性标注

词性标注(Part-Of-Speech Tagging, POS Tagging)也称为语法标注(Grammatical Tagging)或词类消疑(Word-Category Disambiguation),是语料库语言学(Corpus Linguistics)中将语料库内单词的词性按其含义和上下文内容进行标记的文本数据处理技术。

词性标注的目标是用一个单独的标签标记每一个词,该标签表示了用法和其句法作用,如名词、动词、形容词等。

在自然语言分析中,机器需要模拟分析语言。为了达到所需要的效果,机器必须在一定程度上能够理解自然语言的规则。

8.2 关系抽取

若有两个存在着关系的实体,可将两个实体分别称为主体和客体,那么关系抽取就是在非结构或半结构化数据中找出主体与客体之间存在的关系,并将其表示为实体关系三元组,即(主体,关系,客体)。

8.3 信息熵

8.3.1 信息熵的概念

信息熵是一个抽象的概念,可以将信息熵理解为某种特定信息的出现概率。一个系统越是有序,信息熵就越低;反之,则信息熵越高。信息熵同样可以理解为系统有序化程度的一个度量。

信息熵是对信息的度量。对于个人来说,一件事的不确定性越大,相应的信息就越大,对这个信息进行传输或者存储就会消耗更多的能量。

8.3.2 信息熵的计算

一般而言,当一种信息出现概率更高的时候,说明它被传播得更广泛,换言之,被引用的程度更高。因此可以认为,从信息传播的角度来看,信息熵可以表示信息的价值,从而得到一个衡量信息价值高低的标准。

信源的平均不定度。在信息论中信源输出是随机量,其不定度可以用概率分布来度量。记作

$$H(X) = H(P_1, P_2, \cdots, P_n) = -P(x_i)\log P(x_i)$$

这里 $P(x_i)$,$i = 1, 2, \cdots, n$ 为信源取第 i 个符号的概率。$P(x_i)=1$,$H(X)$ 称为信源的信息熵。

熵的概念来源于热力学。在热力学中熵的定义是系统可能状态数的对数值,称为热熵。热力学指出,对任何已知孤立的物理系统的演化,热熵只能增加,不能减少。但这里的信息熵则相反,它只能减少,不能增加。任何系统要获得信息必须要增加热熵来补偿,因此两者在数量上是有联系的。

可以从数学上加以证明，只要 $H(X)$ 满足下列三个条件：

① 连续性：$H(P, 1-P)$ 是 P 的连续函数（$0 \leqslant P \leqslant 1$）；

② 对称性：$H(P_1, \cdots, P_n)$ 与 $P_1, \cdots, P_n$ 的排列次序无关；

③ 可加性：若 $P_n = Q_1 + Q_2 > 0$，且 $Q_1, Q_2 \geqslant 0$，则有 $H(P_1, \cdots, P_{n-1}, Q_1, Q_2) = H(P_1, \cdots, P_{n-1}) + P_n H$；则一定有下列唯一表达形式：

$$H(P_1, \cdots, P_n) = -CP(x_i)\log P(x_i)$$

其中 C 为正整数，一般取 $C = 1$，它是信息熵的最基本表达式。

信息熵的单位与公式中对数的底有关。最常用的是以 2 为底，单位为比特（bit）；在理论推导中常采用以 e 为底，单位为奈特（Nat）；还可以采用其他的底和单位，并可进行互换。

信息熵除了上述三条基本性质外，还具有一系列重要性质，其中最主要的有：

① 非负性：$H(P_1, \cdots, P_n) \geqslant 0$；

② 确定性：$H(1, 0) = H(0, 1) = H(0, 1, 0, \cdots) = 0$；

③ 扩张性：$H_{n-1}(P_1, \cdots, P_n - \varepsilon, \varepsilon) = H_n(P_1, \cdots, P_n)$；

④ 极值性：$P(x_i)\log P(x_i) \leqslant P(x_i)\log Q(x_i)$；这里 $Q(x_i) = 1$；

⑤ 上凸性：$H[\lambda P + (1-\lambda)Q] > \lambda H(P) + (1-\lambda)H(Q)$，式中 $0 < \lambda < 1$。

8.4 新词提取

8.4.1 新词发现

在处理文本对象时，最重要的环节为"分词"，几乎所有的过程及结果都依赖分词。因此，分词的准确性在很大程度上影响着后续的处理，分词结果的不同，也就影响了特征的提取。

新词发现是通过对已有语料进行挖掘，从中识别出新词。新词是指随时代发展而新出现或旧词新用的词语。

8.4.2 短语提取

短语提取是指从文章中提取典型的、有代表性的短语，期望能够表达文章的关键内容。关键短语抽取对于文章理解、搜索、分类、聚类都很重要。而高质量的关键短语抽取算法，还能有效助力构建知识图谱。

8.4.3 新词提取

可以将新词提取分为两部分：新词发现和短语提取。依据某种手段或方法，从文本中挖掘词语，组成新词表。借助挖掘得到的新词表，和之前已有的旧词表进行比对，不在旧词表中的词语即可认为是新词。

新词提取的传统做法为基于分词的方法，现在也慢慢衍生出来基于规则和基于统计的方法。

1. 基于分词的方法

先对文本进行分词，剩下未能被成功分词的部分就是新词。

新词提取的目标是找出文本中的新词，以此来帮助后续的分词等工作。而基于分词的方法则相当于从目标反推过程，这种方法未尝不可行，但分词的准确性本身依赖于词库的完整性，若是词库都不完整，那分词得来的结果也不会准确。

2. 基于规则的方法

根据新词的构词特征或外形特征建立规则库、专业词库或模式库，通过编写正则表达式的方式，从样本中将这些内容抽取出来，并依据相应的规则进行清洗和过滤，然后通过规则匹配发现新词。

这种方法的新词提取的效果非常好，夹杂中文、英文和标点符号的词语也能够被提取。但其与领域耦合过深，无法将建立的规则库、专业词库以及模式库迁移到其他领域中。无论是搭建还是后续的维护，都需要大量的人工成本。

3. 基于统计的方法

基于统计的方法根据有无标注数据可分为有监督方法和无监督方法。

（1）有监督方法。有监督方法利用标注语料，将新词发现看作分类或者序列标注问题。基于文本片段的某些统计量，以此作为特征训练二分类模型；基于序列信息进行序列标注直接得到新词，或对得到的新词再进行判定。

（2）无监督方法。不依赖于任何已有的词库、分词工具和标注语料，只根据词的共同特征，利用统计策略将一段语料中可能成词的文本片段全部提取出来，然后再利用语言知识排除不是新词语的"无用片段"或者计算相关度，寻找相关度最大的字与字的组合。最后，把所有抽取得到的词和已有的词库进行比较，就能得到新词。

8.5 关键词提取和词频统计

8.5.1 关键词提取

关键词是能够表达文本中心的词语。关键词提取是文本挖掘中的一部分，是大部分文本挖掘研究的基础性工作。

从算法的角度来看，关键词提取算法主要分为两种：无监督关键词提取、有监督关键词提取。

1. 无监督关键词提取方法

不需要人工标注的文本，利用一些方法来发现文本中比较重要的词，并将其作为关键词，进行关键词提取。这个方法是先标注出候选词，然后对各个候选词进行打分，并输出一个分值最高的候选词来作为关键词。

2. 有监督关键词提取方法

将关键词抽取过程视为二分类问题，先提取出可能会是关键词的词语，然后对每个选出来的词语进行标签，分为关键词和普通词，并训练关键词抽取分类器。

3. 无监督方法和有监督方法对比

无监督方法不需要人工标注训练的过程，因此速度比较快，但由于无法结合全部文本并利用多种信息对候选词进行排序，所以效果稍逊于有监督方法；而有监督方法可以通过训练来学习如何更好地调节多种信息对判断关键词的影响程度，因此效果更佳。

8.5.2 词频统计

对语篇或语料库中某一语词或短语出现的频数进行统计的过程或结果。词频统计是指输入一些字符串，用程序来统计这些字符串中总共有多少个单词，以及每个单词出现的次数。单词的总数为不重复的单词数总和。

下面结合实践，学习如何实现词频统计。

实践环境：Python 3.9。

实践代码：

```python
import matplotlib.pyplot as plt
import jieba
import wordcloud
import matplotlib

matplotlib.rcParams['font.sans-serif'] = ['simple']   # 设置绘图字体


def wordFreq(filepath, text, topn):
    # Jieba 分词库分词
    words = jieba.lcut(text.strip())
    counts = {}
    # 列表生成式获取停用词
    stopwords = [line.strip() for line in open('stop_words.txt', 'r', encoding='utf-8').readlines()]
    word_clear = []   # 用于生成词云的词语列表，避免重复分词，节约运行时间

    # 统计词频
    for word in words:



        if (len(word) == 1):
            continue
        elif word not in stopwords:
            if word == "王夫人" or word =="凤姐儿":
                word = "凤姐"
            elif word == "林黛玉" or word == "林妹妹" or word == "黛玉笑":
                word == "黛玉"
            elif word == "宝二爷":
                word == "宝玉"
```

```
            elif word == "袭人道":
                word == "袭人"
            elif word == "老太太" or word == "奶奶":
                word = "贾母"
            word_clear.append(word)
            counts[word] = counts.get(word, 0) + 1

    items = list(counts.items())
    items.sort(key=lambda x: x[1], reverse=True)
    for i in range(topn):
        word, count = items[i]
        print(f"{word}:{count}")
    return word_clear


def gen_cloudword(txt):
    wcloud = wordcloud.WordCloud(font_path=r'C:\Windows\Fonts\simhei.ttf',
width=1000, max_words=100, height=860,
                                   margin=2).generate(txt)
    wcloud.to_file("红楼梦cloud_star.png")   # 保存图片
    # 显示词云图片
    plt.imshow(wcloud)
    plt.axis('off')
    plt.show()


text = open('红楼梦.txt', "r", encoding='utf-8').read()
words_clear = wordFreq('红楼梦.txt', text, 10)

gen_cloudword(' '.join(words_clear))
```

词频统计结果如图 8.2 所示。

```
宝玉:3867
凤姐:2652
贾母:2250
贾琏:697
平儿:601
袭人:593
宝钗:571
黛玉:568
老爷:539
丫头:532
```

图8.2　词频统计结果

词云示意图如图 8.3 所示。

图8.3　词云示意图

实战应用一：文本关键词提取

应用背景：在上一单元中解决了命名实体识别问题，接下来继续完成热点问题挖掘模型，涉及的另一个关键环节为关键词提取。解决这一问题后，可以有效支撑整个热点挖掘算法构建。请利用所学知识解决这一问题。

```
# coding=utf-8
from textrank4zh import TextRank4Keyword, TextRank4Sentence
import jieba.analyse
from snownlp import SnowNLP
import pandas as pd
import numpy as np


# 关键词抽取
def keywords_extraction(text):
    tr4w = TextRank4Keyword(allow_speech_tags=['n', 'nr', 'nrfg', 'ns', 'nt', 'nz'])
    # allow_speech_tags   --词性列表，用于过滤某些词性的词
    tr4w.analyze(text=text, window=2, lower=True, vertex_source='all_filters', edge_source='no_stop_words',pagerank_config={'alpha': 0.85, })
```

```
        # text             --  文本内容,字符串
        # window           --  窗口大小,int。用来构造单词之间的边。默认值为2
        # lower            --  是否将英文文本转换为小写,默认值为False
        # vertex_source    --  选择使用words_no_filter, words_no_stop_words, words_
all_filters中的哪一个来构造pagerank对应的图中的节点
        #                  --  默认值为'all_filters',可选值为'no_filter', 'no_stop_
words', 'all_filters'
        # edge_source      --  选择使用words_no_filter, words_no_stop_words, words_
all_filters中的哪一个来构造pagerank对应的图中的节点之间的边
        #                  --  默认值为'no_stop_words',可选值为'no_filter', 'no_stop_
words', 'all_filters'。边的构造要结合'window'参数

        # pagerank_config  --  pagerank算法参数配置,阻尼系数为0.85
        keywords = tr4w.get_keywords(num=6, word_min_len=2)
        # num              --  返回关键词数量
        # word_min_len     --  词的最小长度,默认值为1
        return keywords

    # 关键短语抽取
    def keyphrases_extraction(text):
        tr4w = TextRank4Keyword()
        tr4w.analyze(text=text, window=2, lower=True, vertex_source='all_filters', edge_source='no_stop_words',
                     pagerank_config={'alpha': 0.85, })
        keyphrases = tr4w.get_keyphrases(keywords_num=6, min_occur_num=1)
        # keywords_num     --  抽取的关键词数量
        # min_occur_num    --  关键短语在文中的最少出现次数
        return keyphrases


    # 关键句抽取
    def keysentences_extraction(text):
        tr4s = TextRank4Sentence()
        tr4s.analyze(text, lower=True, source='all_filters')
        # text             --  文本内容,字符串
        # lower            --  是否将英文文本转换为小写,默认值为False
        # source           --  选择使用words_no_filter, words_no_stop_words, words_all_filters
中的哪一个来生成句子之间的相似度
        #                  --  默认值为'all_filters',可选值为'no_filter', 'no_stop_words',
'all_filters'
        # sim_func         --  指定计算句子相似度的函数

        # 获取最重要的num个长度大于等于sentence_min_len的句子用来生成摘要
        keysentences = tr4s.get_key_sentences(num=3, sentence_min_len=6)
```

```python
        return keysentences

def keywords_textrank(text):
    keywords = jieba.analyse.textrank(text, topK=6)
    return keywords


if __name__ == "__main__":
    text = "来源：中国科学报本报讯（记者肖洁）又有一位中国科学家喜获小行星命名殊荣！ 4月19日下午，中国科学院国家天文台在京举行"周又元星"颁授仪式，" \
           "我国天文学家、中国科学院院士周又元的弟子与后辈在欢声笑语中济济一堂。国家天文台党委书记，" \
           "副台长在致辞一开始更是送上白居易的诗句："令公桃李满天下，何须堂前更种花。"" \
           "据介绍，这颗小行星由国家天文台施密特CCD小行星项目组于1997年9月26日发现于兴隆观测站，" \
           "获得国际永久编号第120730号。2018年9月25日，经国家天文台申报，" \
           "国际天文学联合会小天体联合会小天体命名委员会批准，国际天文学联合会《小行星通报》通知国际社会，" \
           "正式将该小行星命名为"周又元星"。"
    # 关键词抽取
    keywords=keywords_extraction(text)
    print(keywords)

    # 关键短语抽取
    keyphrases=keyphrases_extraction(text)
    print(keyphrases)

    # 关键句抽取
    keysentences=keysentences_extraction(text)
    print(keysentences)

    # 基于Jieba的textrank算法实现
    keywords=keywords_textrank(text)
    print(keywords)
```

基于 Jieba 的关键词提取结果：

"小行星" "命名" "国际" "中国" "国家" "天文学家"

感兴趣的同学可以比较两种方法关键词提取的差异，并分析具体原因。

实战应用二：手机评论标签提取

应用背景：随着经济全球化和互联网的快速发展，以及各种智能终端的不断普及，网上购物的热度不断提升。在消费者进行网络购物的同时，也产生了海量的评论数据，而这些评论数据中蕴含着巨大的挖掘价值：对商品厂家来说，评论数据能够直观地反映出用户对商品特性的评价，能够根据用户的喜好调整产品特性，从而更好地发展自身商品；对电商平台来说，可以根据评论数据提取商品标签，提高用户的购物体验，还可以根据用户兴趣进行相关推荐；对用户来说，评论数据是用户了解商品特征的主要信息，用户可以参考评论数据选择自己想要的商品。对用户评论数据进行挖掘，提取出商品标签，可以广泛应用于商品推荐、个性化搜索等场景，有利于商品厂家分析产品数据，有利于提高用户的购物体验，有利于增加平台用户流量。因此，对用户评论数据挖掘进行研究，可以更加有效地提高商品标签的准确性和全面性，在现实生活中，具有十分巨大的价值和深远的意义。

随着公司业务的不断扩大，公司的知名度不断提升，接到了越来越多的广告推介业务。为了保护消费者权益，公司决定对广告合作产品进行消费者评价调研以决定是否对其进行推广，一种比较直观有效的方式为从知名购物网站提取用户评价进行分析。产品经理将这一任务交于你所在的团队，请用所学知识解决。

1. 数据集

爬取某购物网站，获取手机评论标签。

```python
# !/usr/bin/python
# -*-coding:utf-8-*-
# 导入所需的开发模块
import requests
import re
from bs4 import BeautifulSoup,NavigableString
import time
import random

# 创建循环链接
urls = []
for i in list(range(0,6000)):
    urls.append('https://xxx.xxx.xxx'%i)# 替换为需要链接的网站

# 获取链接地址
homepage = 'https://xxx.xxx.xxx'# 替换为需要链接的网站
headers = {'User-Agent': 'Mozilla/5.0 (X11; Ubuntu; Linux x86_64; rv:41.0) Gecko/20100101 Firefox/41.0'}
cookies = requests.get(homepage,headers=headers).cookies

ratecontent = []
```

```
# 循环抓取数据
for url in urls:
#     content=(requests.get(url).text)
    r = requests.get(url,headers=headers,cookies=cookies).text
    ratecontent.extend(re.findall(re.compile('"rateContent":"(.*?)","rateDate
"'),r))    # 正则化提取评论内容
    print(url)
    time.sleep(random.uniform(3,8))

file =open('手机评论标签.csv','w')
for i in ratecontent:
    file.write(i+'\n')
file.close()
```

2. 提取关键词

```
# !/usr/bin/env python3
# -*- coding: utf-8 -*-

import jieba
import jieba.analyse
import logging

logging.basicConfig(format='%(asctime)s : %(levelname)s : %(message)s',
level=logging.INFO)
# 设置日志

content = open('/mnt/share/jieba_test/手机评论便签.csv','rb').read()

tagsA = jieba.analyse.extract_tags(content, topK=20,allowPOS='a')
# allowPOS 是选择提取的词性，a 是形容词

tagsN = jieba.analyse.extract_tags(content, topK=20, allowPOS='n')
# allowPOS='n'，提取名词
```

提取结果如图 8.4 所示。

tagsA2
'不错,流畅,清晰,便宜,漂亮,合适,耐用,很棒,太慢,惊讶,方便,正好,顺畅,耐心,很轻,挺好用,完美,很好,舒服,惊艳,'

图8.4　提取结果

3. 制作预料库

```
import pandas as pd
import numpy as np
import logging
```

```
import codecs

words=jieba.lcut(content,cut_all=False)    # 分词,精确模式

# 去停用词,先制作好"停用词表.txt"
stopwords = []
for word in open("/mnt/share/jieba_test/stopword.txt", "r"):
    stopwords.append(word.strip())
stayed_line = ""
for word in words:
    if word not in stopwords:
        stayed_line += word + " "

# 保存语料
file=open('/mnt/share/jieba_test/corpus.txt','wb')
file.write(stayed_line.encode("utf-8"))
file.close()
```

分词后语料如下所示:

速度很快手感一流系统流畅稳定大爱送父母换送荣耀P10 肯定体验更好水果退休哈哈哈 颜值超级高滑背面流光溢彩真的太好看帮朋友买说王者荣耀一点不卡充电不发烫充电速度 荣耀手感不错处理器中高端照相机很棒颜值不赖
手感不错好看信号流畅5.5屏会更好
国产大牌手感舒服颜色不错做工精细屏显清晰细腻艳丽摄像清晰功能强大五星好评 挺不错手机喜欢荣耀品牌国产机中代表放心信号续航长设计新质量棒
手机收到特意几天评价玩游戏手感满意日常第一次带有指纹识别手机总的来说体验不错还会
国产手机特地几天评价速度手感不错照相特清晰支持
帮一位同事购买同事签收前几天问这位同事说好用颜值高喜欢上网很快去年十一买女儿入手荣耀样子小点女生手感更好小巧精致玻璃屏有质感显档次 外观漂亮运行速度国产荣耀系列
特意几天评价4G运行内存安装点软件流畅
蓝色手机棒外观漂亮绚丽第二部荣耀手机大爱操作灵敏运行特别完美
10再来评价手机外观手感速度很快通话信号质量充电电池容量很大满意购物买第三台
适合玩游戏
好用不卡顿流畅王者荣耀没得说值得购买赶紧买
超爱大小合适拍照效果想要
荣耀手机外观挺漂亮运行状态不错
机器漂亮送两个原厂壳提个建议手机充电器做漂亮
宝贝很漂亮屏够用大点更好还会关注荣耀
手机太棒妈妈买老妈手机内存运行慢换新手机满意款荣耀性价比赞客服象服务态度 两周手机手感棒速度流畅外观美丽大气大小合适真的非常喜欢
不错 机子 外形 喜欢

4. word2vec 训练评论语料

```
from gensim.models import word2vec
```

```
sentences = word2vec.Text8Corpus('/mnt/share/jieba_test/corpus.txt')   # 加载
制作好的语料
model = word2vec.Word2Vec(sentences, size=200)   # 默认 window=5

commit_index=pd.DataFrame(columns=['commit','similarity'],index=np.arange(100))

index=0
for i in tagsN:
    for j in tagsA:
        commit_index.loc[index,:]=[i+j,model.similarity(i,j)]
        index+=1

comit_index_final=commit_index.sort(columns='similarity',ascending=False)
comit_index_final.index=commit_index.index
```

最终的提取结果见表 8.1。

表8.1 提取结果

index	commit	similarity
1	机子漂亮	0.968
2	指纹方便	0.6918
3	指纹太慢	0.891
4	颜色漂亮	0.867
5	粉色便宜	0.832
6	客服耐心	0.817
7	价格合适	0.814
8	发货太慢	0.779
9	客服很棒	0.765
10	发货方便	0.745
11	机子精致	0.722
12	颜色正好	0.688
13	价格不错	0.681
14	手机便宜	0.662
15	手机顺畅	0.658
16	颜色精致	0.657
17	收货完美	0.646
18	棒棒耐心	0.645
19	颜色惊艳	0.628

单元小结

自然语言处理中的各类问题都可以抽象为：如何从形式与意义的多对多映射中，根据语境选择一种映射。需要运用知识来约束，才能让机器解决形式和意义的多对多映射的困境，而信息提取就是知识获取的一种重要手段。

本单元主要阐述了信息抽取任务的主要方法和信息抽取的基础知识，并通过实战任务解决商品评论分析任务。

下一单元将学习文本聚类任务，对提取的关键信息进行聚类分析。

习 题

1. 什么是信息抽取？
2. 什么是关键词提取和词频统计？它们在文本处理中有何用途？

单元9 文本聚类

通过上一单元的学习,能够从海量数据中提取关键的信息,但是仅仅做到这一步还不够,需要对提取到的信息进行进一步分析——对数据进行分堆(分类处理),以便进一步筛选,这就涉及文本聚类任务。

本单元将学习一个简单但重要的应用:文本聚类。聚类算法是一类速度快、不用标注数据的无监督学习方法。本单元主要专注于聚类算法在文本数据上的应用,学习完本单元后,同学们将能处理一些简单的文本聚类问题,并为进一步学习文本分类方法打下坚实基础。本单元知识导图如图9.1所示。

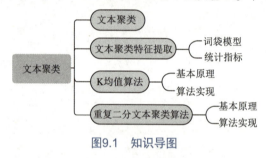

图9.1 知识导图

课程安排

课程任务	课程目标	安排学时
了解文本聚类概念	熟悉文本聚类概念	1
了解文本聚类特征提取	了解词袋模型及词袋模型统计指标	1
掌握k均值算法	掌握k均值算法概念及简易实现方法	1
掌握重复二分文本聚类算法	掌握算法的原理及算法的代码实现	1
实战应用	通过该实战应用来引导同学们了解一些业务背景,解决实际问题,进行自我练习、自我提高	2

9.1 文本聚类概述

要介绍文本聚类,首先就要介绍聚类的概念。物以类聚、人以群分。只要细心观察,会发现日常生活中有许多行为都蕴含着聚类的思想,即使你事先不明白聚类的概念,也能将其运用得灵活自如。例如,如图9.2所示,假设现在草原上有100只羊,牧羊人会通过观察每只羊是否有角、羊毛的卷曲程度、羊蹄的形状将其划分为山羊和绵羊两类。这就是知识来源于生活的一个鲜活例子，事实上许多人工智能的算法都来自于实际生活或自然界。

下面给聚类一个形式化的定义：将包含多个对象的一个集合根据对象特征划分为若干个子集的过程就是聚类。每一个子集也被称为簇，而聚类的总体目标就是让簇内元素的相似度最大，而让簇间元素的相似度最小。本单元将介绍一种文档层级上的聚类任务，即文本聚类。

文本聚类（Text Clustering）也称为文档聚类（Document Clustering），它本质上和前述羊聚类的例子没有什么不同，只是聚类的对象是文本文档而已。而且其作为一种无监督的模型，无须训练也无须预先对文档进行类别标注，十分灵活。目前已经成为对文本信息进行有效组织、整理和分析的重要手段。

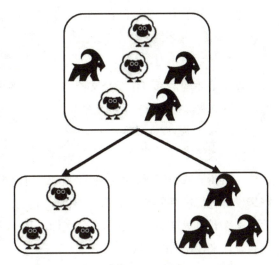

图9.2　羊聚类

文本聚类的基本流程可分为特征提取和向量聚类两步。只有对数值型向量才能进行聚类操作，而很难直接对文本进行聚类。所以想要实现文本聚类，首先需要提取文档的特征，将其转化为文档向量，才能进一步操作。而一旦将文档表示为向量，剩下的算法就与文档无关了。这种抽象思维无论是从软件工程的角度，还是从数学应用的角度，都十分简洁有效。

9.2　文本聚类特征提取

用计算机对文本进行建模时存在一个问题，即"混乱"，由于人工智能算法的输入/输出通常都是固定长度的数据，而文本或文档中的词数是不定长的。因此包括聚类在内的人工智能算法无法直接处理纯文本，只有在将文本转化为固定长度的文档向量才能进一步操作。而文本向量化的方法有独热编码、词袋模型、word embedding 等。本节主要介绍经典的词袋模型。

9.2.1　词袋模型

词袋模型是简单而流行的一种将文本转化成向量的方式，且十分容易实现。词袋模型包含以下两种信息：

（1）一个已知所有单词的词典。
（2）一种已知单词的评价指标。

本模型之所以被称为词袋，是它只关心单词是否出现，而忽略文档中的单词顺序和结构信息，因此这是一种对文档信息的有损压缩。而在词典之外的词都被记为 OOV（Out-of-Vocabulary）。

就人直观的感觉而言，如果文档具有相似的内容，那么它们就是相似的。

图 9.3 是一个词袋模型的样例，下面具体的解释词袋模型。

图9.3　词袋模型示例

在这个例子中，假设语料库中含有以下两句话：
"我去水里游泳。"
"水好凉！"

将上面每句话都当作一个文档，目标是将每个文档分别表示为一个定长向量。首先，这两句话经过分词和去除标点符号后的结果为：

我 \ 去 \ 水 \ 里 \ 游泳
水 \ 好 \ 凉

一共得到八个词语(共七种)，那么这七种词就是上述语料库所构成的字典。将这七种词语使用词袋模型进行建模。接下来，以一种简单的方式着手构造文档向量。由于得到的词典大小为 7，因此使用一个长度为 7 的向量表示每个文档。其中，向量的每一维分别表示对应位置单词的评分。使用词频统计来计算文档向量。

以第 1 个文档"我去水里游泳。"为例，有：

我 =1
去 =1
水 =1
里 =1
游泳 =1
好 =0
凉 =0

因此，第 1 个文档对应的向量为 [1,1,1,1,1,0,0]。同理，第 2 个文档生成的向量为 [0,0,1,0,0,1,1]。而对于与已知单词的词汇表重叠但可能包含词汇表之外的单词的新文档，仍然可以进行编码，只对已知单词的出现进行评分，忽略未知单词。比如，"冰水好凉"这个文档中，"冰"是一个未被收录进词典的词（OOV），会被词袋模型忽略，其生成的文档向量仍为 [0,0,1,0,0,1,1]。

词袋模型完全不考虑词语的顺序和文档结构的压缩方式，比如对于这个模型来说，"水好凉"和"凉水好"的文档向量是一模一样的。但也正因为此，它的计算成本非常低，所以在实际工程应用中，它仍是一个广泛使用的基线模型。

9.2.2 词袋模型中的统计指标

词袋模型有许多种生成文档向量的统计指标。

（1）布尔词频：布尔变量即 0-1 变量，词频非零向量对应维为 1，否则为 0。

（2）词向量：其思想是将每个词转换为一个向量，然后文档向量就是其中每个词的向量之和或池化等运算。比较经典的方法有 word2vec、GloVe 预训练词向量。词向量中蕴含了词语本身的语义，比如"树"和"叶"的词向量就比"树"和"天"的词向量相似度更高。

下面在图 9.2 的词袋模型中加入"我去水里游泳划水"这句话进行解析。这里假设仍使用原词典，即"划"是 OOV 词。

分词：我 \ 去 \ 水 \ 里 \ 游泳 \ 划 \ 水

词频：我 =1，去 =1，水 =2，里 =1，游泳 =1，好 =0，凉 =0。对应的文档向量为 [1,1,2,1,1,0,0]。

布尔词频：我 =1，去 =1，水 =1，里 =1，游泳 =1，好 =0，凉 =0。对应的文档向量为 [1,1,1,1,1,0,0]。

词向量：假设使用 128 维的词向量，每个词对应的归一化词向量为 $[d_1,d_2\cdots d_{128}]$，那么这个文档向量为每个词向量的拼接：$[\,[d_1,d_2,\cdots,d_{128}]^{我}, [d_1,d_2,\cdots,d_{128}]^{去}, [d_1,d_2,\cdots,d_{128}]^{水}, [d_1,d_2,\cdots,d_{128}]^{里}, [d_1,d_2,\cdots,d_{128}]^{游泳}, [d_1,d_2,\cdots,d_{128}]^{划}, [d_1,d_2,\cdots,d_{128}]^{水}\,]$。

这些方法的效果与实际数据集相关。通常词向量的效果要比词频和布尔词频好，但其成本更高。实际应用时需要权衡各方条件。

词袋模型统计指标见表 9.1。

表9.1 词袋模型统计指标

指　　标	特　　点	使用场景
布尔词频	计算极简单	适合长度较短的数据集
词频	计算简单	适合文档主题较多的数据集
词向量	计算复杂	适合OOV较多的数据集

9.3 k均值算法

在得到文档向量之后，就可以把这些向量视为普通数值向量套用聚类算法了。常用的聚类算法有基于划分的 k 均值算法、基于层次的孤立森林算法、基于网格的 STING 算法和基于密度的 DBSCAN 算法等。

k 均值 (k-means) 算法是一种原理简洁且实用的聚类算法，由 Stuart Lloyd 于 1957 年提出，并在此后的几十年中被广泛使用。

9.3.1 基本原理

为了让同学们更好地理解该模型，下面首先介绍一个关于 k 均值算法的例子：

某街道分到了三台便民医疗车，一开始每辆车随意选了一个地点，并将这三个地点的信息公告给了街道内的所有居民，于是每位居民前去离家最近的地点做健康监测。一周之后，大家觉得距离太远了，于是每个地点的工作人员统计了一下最近一周来检测的所有居民的地址，然后搬到了所有地址的中心地带，并通过广播的方式告知了居民新检测点的位置。医疗车每次移动不可能离所有人都更近，有人发现 1 号车移动后不如去 2 号车更近，于是每位居民又去了离自己最近的检测点……就这样，医疗车每周更新自己的位置，居民根据自己与检测点之间的相对距离选择最近的监测点，最终稳定了下来。

假设现在有 n 个文档向量 $x_1,x_2\cdots,x_n$ 和一个整数 k，k 均值算法的步骤如下：

（1）选择初始化的 k 个文档向量作为初始聚类中心 $a=a_1,a_2,\cdots,a_k$；

（2）针对数据集中每个样本 x_i 计算它到 k 个聚类中心的距离并将其分到距离最小的聚类中心所对应的类中；

（3）针对每个类别 C_i（$1 \leqslant i \leqslant k$）和该类别中的文档向量 x，重新计算它的聚类中心 $a_i = \frac{1}{|C_i|}\sum_{x\in C_i}x$

（即属于该类的所有样本的质心）；

（4）重复步骤（2）和（3），直到达到最大迭代次数或聚类中心不再变化。

9.3.2　k 均值算法的简易实现

为了方便调用，将 k 均指算法包装成一个类，在类初始化时需要指定算法的 k 值和最大迭代次数 maxIter。

```python
import numpy as np

class Kmeans:
    def __init__(self, k, maxIter) -> None:
        """ 构造函数
        k: 聚类的个数
        maxIter: 最大迭代次数
        """
        self.k = k
        self.maxIter = maxIter
```

根据 9.3.1 节中的基本步骤进行聚类部分的编码，这里的结束条件是达到最大迭代次数。最终聚类的结果保存在 self.labels 中，具体过程见注释。

```python
def fit(self, X):
    """
    X: 类数组类型，待聚类的文档向量
    """
    X = np.asarray(X)
    np.random.seed(0)
    # 从数据中随机选出 k 个样本作为初始聚类中心
    self.cluster_centers = X[np.random.randint(0, len(X), self.k)]
    self.labels_ = np.zeros(len(X))

    for t in range(self.maxIter):
        for index, x in enumerate(X):
            dis = np.sqrt(np.sum(x-self.cluster_centers)**2, axis=1)
            # 将最小距离的所有赋值给标签数组，索引的值就说当前点所属的簇
            self.labels_[index] = dis.argmin()
        # 循环遍历每个簇
        for i in range(self.k):
            # 计算每个簇内的所有样本均值，更新聚类中心
            self.cluster_centers[i] = np.mean(X[self.labels_ == i], axis=0)
```

在聚类完成后，如果有一个新的文档需要判断属于哪一类，那么就可调用下面的 predict 函数进行分类。判断准则是看其离哪个簇中心最近，返回值 result 是新文档所属的簇号。

```
def predict(self, X):
    # 预测新文档向量所属的簇
    X = np.asarray(X)
    result = np.zeros(len(X))
    for index, x in enumerate(X):
        # 计算样本到每个聚类中心的距离
        dis = np.sqrt(np.sum(x-self.cluster_centers)**2, axis=1)
        result[index] = dis.argmin()
    return result
```

9.4 重复二分聚类算法

9.4.1 基本原理

重复二分聚类算法是 k 均值算法的效率加强版,它是反复对子集进行二分。该算法的步骤如下:
(1)按照指定规则选择某个簇进行划分。
(2)利用 k 均值算法 (k=2) 将该簇二分为两个子集。
(3)检查目前生成的簇数量是否达到预设数量,若不满足则重复执行前两步,否则算法结束。

回到 9.2 节中羊聚类的例子,假设现在除了山羊和绵羊之外又多了一种羚羊,那么重复二分聚类算法的建模过程如图 9.4 所示。

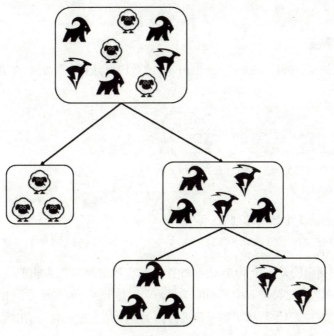

图9.4 重复二分聚类示意图

第一轮迭代时，区分出了绵羊和羚羊、山羊；第二轮迭代时，进一步区分羚羊和山羊。每次迭代都会将某个簇一分为二，形成了图 9.4 中的树结构。从这点看，重复二分聚类算法其实与孤立森林这种层次聚类算法更加相似。但每次划分都基于 k 均值聚类算法 (k=2)，由于每次聚类都仅仅在一个子集上进行，输入数据少，因此算法速度更快。其缺点是对数据输入的顺序敏感，且噪声和离群点对最终结果干扰较大。

而至于每次挑选哪个簇进行划分存在多种方案，比如可每次选最大的簇或选最"稀疏"的簇等，具体需要根据实际情况判断。

9.4.2 算法实现

下面结合一个案例，利用重复二分聚类实现文本相似度检测。

在实际使用 HanLP 计算中文文本相似度时，不需要显式地手动进行特征提取，HanLP 内部会自动进行计算。这种高度的集成性也是 HanLP 广受欢迎的一个重要原因，大大减少了学习使用成本。为了简便起见，此处使用轻量级的 RESTful API。它的大小仅为几千字节，适合敏捷开发、移动 App、课堂教学等场景，简单易用，无须 GPU 配置环境。

1. 安装 RESTful API

```
pip install hanlp_restful
```

在 API 安装成功后，创建 HanLP 客户端，填入服务器地址和密钥：

```
from hanlp_restful import HanLPClient
HanLP = HanLPClient('https://www.hanlp.com/api', auth=None, language='zh')
# auth 不填则为匿名，zh 代表中文
```

2. 相似度检查实战

直接调用客户端的 semantic_textual_similarity( ) 函数即可计算中文文本的相似度。下面计算三对句子的相似度：

```
HanLP.semantic_textual_similarity([
    ('我去水里游泳。', '我去泳池游泳。'),
    ('绵羊、山羊和羚羊都是羊', '羚羊、山羊和绵羊都是羊'),
    ('我叫他去打扫卫生', '他叫我去打扫卫生'),
])
```

输出结果为一个包含 3 个数字的列表：

```
[0.6528788208961487, 0.98851156234744121, 0.0222462117671966655]
```

HanLP 计算出的相似度是一个 [0,1] 之间的浮点数，数值越大代表相似度越高。可以看到"绵羊、山羊和羚羊都是羊"和"羚羊、山羊和绵羊都是羊"是上述三对句子中最相似的句子，相似度高达 0.99；而"我叫他去打扫卫生"和"他叫我去打扫卫生"是上述三对句子中最不相似的句子，相似度仅有 0.02。

实战应用一：食品安全评论聚类

应用背景：食品安全是社会各界日益关注的民生问题，政府部门正在逐步完善监管体制、加大监管力度，构建社会共治的格局。通过分析食品安全评论数据，可以查看差评有没有聚集性。如果差评评价的都是一个商家，那么这个商家可能存在严重的食品安全问题；如果是一个消费者或者一群有关系的消费者，那么这个可能是恶意差评。因此，有了大规模的文本挖掘，就可以发现平台中隐藏的问题，作为感知或者原因归因。

由于团队业绩出色，公司决定组织员工外出团建。团队负责人希望在选定就餐地点时能够避免食品安全风险，希望通过算法进行数据筛选挖掘筛选出相对较好的餐厅。

1. 数据集

选用食品安全评论数据集，数据格式如图 9.5 所示。

```
一如既往地好吃，希望可以开到其他城市
味道很不错，分量足，客人很多，满意
少送一个牛肉汉堡，而且也不好吃，特别是鸡肉卷，都不想评论了，谁买谁知道
环境很好，服务很热情，味道非常好，鱼也很新鲜
一如既往地好吃，个人更喜欢吃全翅
蛋很好吃，送货及时，服务一流，下次有机会再买
一个咸一个酸
价格实惠，服务态度很好，分量超足，鸭爪入口即化，软糯，很入味，超辣，推荐！
菜品多，味道也不错，最好点个鸳鸯锅，锅底也是20元
真的很不错，朋友从外地来，特意带她们来吃的，特别喜欢这个地方，特别是老火锅
很好吃，强力推荐，香菜牛肉果然是招牌菜，名不虚传
味道不错，份量不足
上两次还觉得可以，今天这个菠萝包里头奶油都坏了，完全不能吃
第二次来吃了，味道还可以，就在商场里，逛累了可以来吃
饭里面居然还有虫？不想吃了！差评！
```

图9.5 数据格式

2. 数据处理

（1）数据读取。

```python
# 加载所需要的包
from sklearn.feature_extraction.text import TfidfVectorizer
import pandas as pd
import os
import jieba
# 设置工作空间 & 读取数据
os.chdir('/DataSets/comment_data')
train = pd.read_csv('train.csv',sep='\t')
```

(2)文本向量化。

将文本转换成向量。

```
# 进行分字处理 用结巴分词
data = train['comment'].apply(lambda x:' '.join(jieba.lcut(x)) )

# 进行向量转化
vectorizer_word = TfidfVectorizer(max_features=800000,
                                  token_pattern=r"(?u)\b\w+\b",
                                  min_df = 5,
                                  analyzer='word',
                                  ngram_range=(1,2)
                                  )
vectorizer_word = vectorizer_word.fit(data)
tfidf_matrix    = vectorizer_word.transform(data)
```

3. 模型训练

```
# 设置下最大可展示的行
pd.set_option('display.max_rows', 1000)

# 进行模型训练
from sklearn.cluster import DBSCAN

clustering = DBSCAN(eps=0.95, min_samples=5).fit(tfidf_matrix)
```

可以调整两个主要参数，获取不同的聚类效果。

4. 模型测试

```
# 分群标签打进原来的数据
train['labels_'] = clustering.labels_

# 抽取编号为 5 的群，可以看到，这个群都是关于吃了拉肚子的反馈
for i in train[train['labels_']==5]['comment']:
    print(i)
```

吃了拉肚子。一点当地味道都没有
吃了拉肚子 弄得不卫生
吃了拉肚子 特别不卫生
吃了拉肚子，确实卫生不好
我吃了拉肚子
吃了拉肚子
不干净，吃了拉肚子，还发烧了
老样子，吃了拉肚子
鱼不新鲜，吃了拉肚子
吃了拉肚子，一上午都在医院

```
好像拉肚子了
拉肚子
太辣，吃了拉肚子
吃了拉肚子，现在都十二点多了，还疼得睡不着
我吃了拉肚子
吃了拉肚子
不是正宗味道。不新鲜，吃了拉肚子
泡椒猪肝吃了拉肚子
吃了拉肚子，而且很严重
吃了拉肚子，口感不好
味道不行 吃了拉肚子
别买，不卫生
吃了拉肚子
菜品不新鲜，吃了拉肚子
鸭脚变味了，吃了拉肚子
吃了拉肚子，有点不新鲜了
就是不知道怎么回事
我吃了拉肚子
吃了拉肚子，味道怪怪的
吃了拉肚子
```

下面是比较好评的评价。

```
# 抽取编号为5的群，可以看到，这个群都是好评的
for i in train[train['labels_']==5]['comment']:
    print(i)
```

```
上菜快，味道不错
环境还不错，上菜速度快，服务态度好
环境好，上菜速度快，味道特别好吃，去吃了很多次了
菜品很不错，上菜速度快，环境不错
便宜，味道好，上菜快，环境好，菜品丰富不错，环境好，上菜速度快
还不错，上菜速度快，口味也合适，可以试一试，经常去
味道好，上菜速度快，环境也不错，口味不错，上菜速度快，推荐品尝
菜品新鲜，味道不错，上菜速度快
经常去，味道不错，上菜速度快。服务很好，上菜速度快。味道不错。环境很好，上菜速度快。下次还来。很好，速度快，环境也不错
```

实战应用二：汽车竞品分析

应用背景：公司汽车专栏准备组织一期新能源汽车专栏，目前市场上的新能源汽车品牌种类繁多。公司希望以竞品分析为背景，通过数据的聚类，为汽车提供聚类分类。对于指定的车型，可以通过聚类分析找到其竞品车型。利用车型数据，进行车型画像的分析，为产品的定位，竞品分析提供数据决策。

1. 数据集

数据格式见表9.2。

表9.2 数据示意

symboling	保险风险等级：数越小，越安全
carComany	车辆公司
fueltype	燃料类型
aspiration	送气
doornumber	门数量
carbody	车身
drivewheel	驱动轮
wheel base	轴距
carlength	车长
canwidth	车宽
carheight	车高
curbweight	空车重
enginetype	发动机类型
cylindernumber	气缸编号
enginesize	车辆尺寸
fuelsystem	车辆燃料类型
stroke	发动机内冲程或容积
compressionratio	压缩
peakrpm	峰值转速
citympg	城市里程

2. 数据处理

（1）数据导入。

```
import pandas as pd
import numpy as np
import matplotlib.pyplt as plt
car_data=pd.read_csv('./car_price.csv')
car_data.head()
```

（2）去除重复项。

```
car_data.info()# 查看26个字段的类型
car_data.duplicated().sum()# 查看是否有重复项

# 调取每个类型下(int 64)的数据，并进行描述统计
car_data_int=car_data.select_dtypes(include='int64')
car_data_int.describe()# 根据结果显示int类型的数据正常，没有异常值
```

```python
# 调取每个数值类型（float）下的数据，并进行描述统计
car_data_float=car_data.select_dtypes(include='float64')
car_data_float.describe()# 根据结果显示float64类型的数据正常，没有异常值

# 调取文本类型下的数据，并进行描述统计，查看文本值的分布情况
car_data_object=car_data.select_dtypes(include='object')
for i in car_data_object.columns:
    print(i)
    print(set(car_data_object[i]))# set是一组dict字典形式，可以过滤重复值

# 将潜在的数值型文本数据进行转化
object_dic={'two':2,'four':4, 'six':6, 'eight':8, 'twelve':12, 'three':3, 'five':5, 'four':4}
car_data['doornumber']=car_data['doornumber'].replace(object_dic)
car_data['cylindernumber']=car_data['cylindernumber'].replace(object_dic)

# 对品牌名称进行处理
car_data['CarName']=car_data['CarName'].apply(lambda x :str(x).split(' ')[0])
car_data.head()# 注意用str(x)而不是x.split() 否则报错
```

检查拼写错误并替换：

```python
# 查看有哪些品牌名
car_data_name=set(car_data['CarName'])
car_data_name

# 改正拼写及大小写错误
car_data_name={'maxda':'mazda','Nissan':'nissan','porsche':'porcshce','toyouta':'toyota','vokswagen':'volkswagen','vw':'volkswagen'}# 字典形式
car_data['CarName']=car_data['CarName'].replace(car_data_name)
car_data_name2=set(car_data['CarName'])
car_data_name2
```

更改后的品牌名称如下：

```
'alfa-romero',
'audi',
'bmw',
'buick',
'chevrolet',
'dodge',
'honda',
'isuzu',
'jaguar',
'mazda',
```

```
'mercury',
'mitsubishi',
'nissan',
'peugeot',
'plymouth',
'porcshce',
'renault',
'saab',
'subaru',
'toyota',
'volkswagen' ,
'volvo'
```

3. 数据分析

(1)统计图分析。

```
car_data_float=car_data.select_dtypes(include='float64')
car_data_float.describe()

num_cols=car_data_float.columns
import seaborn as sns
fig=plt.figure(figsize=(12,8))
i=1

  for col in num_cols:
    ax=fig.add_subplot(2,4,i)
    sns.boxplot(data=car_data[col],ax=ax,sym="r+")
    i=i+1
    plt.title(col)
plt.show()
```

结合描述性统计和箱性统计,确定是否有异常值。

(2)相关性分析。

```
# 对数值型进行相关性分析,corr只对数值型进行操作
car_data_corr=car_data.corr()
car_data_corr
```

为了更直观地观察数据间的相关性,可以绘制热力图:

```
# 使用mask只绘制热力图其中的一半
mask = np.zeros_like(car_data_corr)
mask[np.triu_indices_from(mask)] = True
with sns.axes_style("white"):
        f, ax = plt.subplots(figsize=(25, 10))
```

```
        ax = sns.heatmap(car_data_corr,annot = True, mask=mask,vmin=0,vmax=1,
square=True)
    ax.set_title('correlation between car')
```

0.8～1.0 表示极强相关；0.6～0.8 表示强相关；0.4～0.6 表示中等程度相关；0.2～0.4 表示弱相关；0.0～0.2 表示极弱相关或无相关。

（3）分箱操作。

根据车长、轴型可以分为六类：

微型车（A00）：车长小于 3.7 m；轴距小于 2.35 m；

小型车（A0）：车长小于 4.3 m；轴距小于 2.5 m；

紧凑型车（A）：车长小于 4.6 m；轴距小于 2.7 m；

中型车（B）：车长小于 4.9 m；轴距小于 2.8 m；

中大型车（C）：车长小于 5.1 m；轴距小于 2.9 m；

豪华车（D）：车长大于 5.1 m；轴距大于 2.9 m。

根据分类标准进行分箱操作：

```
bin=[min(car_data.carlength)-0.01,145.67,169.29,181.10,192.91,200.79,max(car_data.carlength)+0.01]# 确定分组标准
groupnames=['A00','A0','A','B','C','D']# 定义箱名
car_group=pd.cut(car_data.carlength,bin,labels=groupnames)

# 新建一列将分组情况放入数据集
car_data['cargroup']=car_group
car_data.head()
```

4. 聚类处理

（1）特征处理。

```
cardata=car_data.copy()
cardata.head()
cardata['cargroup']=cardata['cargroup'].astype(str)
# 进行编码之前将 cargroup 转换成字符型，否则报错
from sklearn.preprocessing import LabelEncoder
cargroup_copy=LabelEncoder().fit_transform(cardata['cargroup'])
cardata['cargroup']=cargroup_copy

# 剔除车长度，用车型代替（车型是根据车长得到的）
cardata=cardata.drop(['carlenrth'],axis=1)
cardata_features=cardata.select_dtypes(include='object')
cardata_features.head()

# 进行 one-hot 编码
cardata_onehot=pd.get_dummies(cardata_features,drop_first='True')# drop_first
去除冗余特征，避免特征重复
```

```python
cardata2=cardata.join(cardata_onehot).drop(cardata_features,axis=1)# 把处理后的数据加上，处理前的数据删除
cardata2.info()
```

（2）标准化。

```python
from sklearn.preprocessing import MinMaxScaler
cardata2=MinMaxScaler().fit_transform(cardata2)
cardata2.shape

# 将数据转化成矩阵形式
cardata2=pd.DataFrame(cardata2)
cardata2.head()
```

（3）降维。

```python
from sklearn.decomposition import PCA
pca=PCA(.99)# 可解释方差比例设为0.99
cardata3=pca.fit_transform(cardata2)
ratio=pca.explained_variance_ratio_
print('各主成分的解释方差占比：', ratio)
print('降维后有几个成分：', pca.n_components_)
cum_ratio=np.cumsum(ratio)
print('累计解释方差占比：',cum_ratio)

fig=plt.figure(figsize=(10,5))
X=range(1,len(ratio)+1)
plt.bar(X,ratio,edgecolor='blue')
plt.plot(X,cum_ratio,'r.-')
plt.xlabel('PCA')
plt.ylabel('explained_variance_ratio_')
plt.grid(b="True",axis="y")
plt.grid(b="True",axis="x")
plt.show()
```

（4）聚类。

```python
# 确定聚类个数
from sklearn.cluster import KMeans
sse=[]
kvals=range(1,20)# 20是自己设置的，确保明显地看到拐点
for k in kvals:
    Kmeans=KMeans(n_clusters=k,init='k-means++',n_init=10,max_iter=300,random_state=0)# 构造聚类器
    Kmeans.fit(cardata4)# 放入数据
    sse.append(Kmeans.inertia_)
X=kvals
```

```python
plt.xlabel('k')
plt.ylabel('sse')
plt.plot(X,sse,'o-')
plt.grid(b="True",axis="y")
plt.grid(b="True",axis="x")
plt.show()

# k-means 聚类
kms=KMeans(n_clusters=5,init='k-means++',n_init=10,max_iter=300,random_state=123)
# random_state 没有特殊含义，保证每次运行代码选取的中心点一致
# n_clusters 聚类个数为 5 的情况下聚类效果最好

kms.fit(cardata4)
label=kms.labels_ # 获取聚类结果
print(label)

# 绘制散点图
fig=plt.figure(figsize=(8,8))
plt.scatter(cardata4[label==0].iloc[:,0],cardata4[label==0].iloc[:,1],c="red",marker='o',label='class0') # 红色圆圈样式绘制散点图
plt.scatter(cardata4[label==1].iloc[:,0],cardata4[label==1].iloc[:,1],c="red",marker='+',label='class1') # 红色十字样式绘制散点图
plt.scatter(cardata4[label==2].iloc[:,0],cardata4[label==2].iloc[:,1],c="green",marker='o',label='class2') # 绿色圆圈样式绘制散点图
plt.scatter(cardata4[label==3].iloc[:,0],cardata4[label==3].iloc[:,1],c="green",marker='+',label='class3') # 绿色十字样式绘制散点图
plt.scatter(cardata4[label==4].iloc[:,0],cardata4[label==4].iloc[:,1],c="blue",marker='o',label='class4') # 蓝色圆圈样式绘制散点图
plt.scatter(cardata4[label==5].iloc[:,0],cardata4[label==5].iloc[:,1],c="blue",marker='+',label='class5') # 蓝色十字样式绘制散点图
plt.xlabel('pc1')
plt.ylabel('pc2')
plt.title('K-Means PCA')
plt.show()
```

聚类结果分析：

```python
cardata_km=car_data.copy()
cardata_km['km_result']=label
# 统计每个集群，每个车品牌的车型数
cardata_km_num2=cardata_km.groupby(by=['km_result','CarName'])['car_ID'].count()
cardata_km_num2

# 指定类别车型分析
cardata_km_result0=cardata_km.loc[cardata_km['km_result']==0]
```

```
cardata_km_result1=cardata_km.loc[cardata_km['km_result']==1]
cardata_km_result4=cardata_km.loc[cardata_km['km_result']==4]
cardata_km_result=pd.concat([cardata_km_result0,cardata_km_result1,cardata_
km_result4],axis=0)# pd.concat 将数据合并，行向合并

# A 车型占比最大，故将 A 车型下指定的品牌作为竞品
dfa=df.loc[df['cargroup']=='A']
df_com=set(dfa['CarName'])
df_com
```

聚类分析结果如图 9.6 所示。

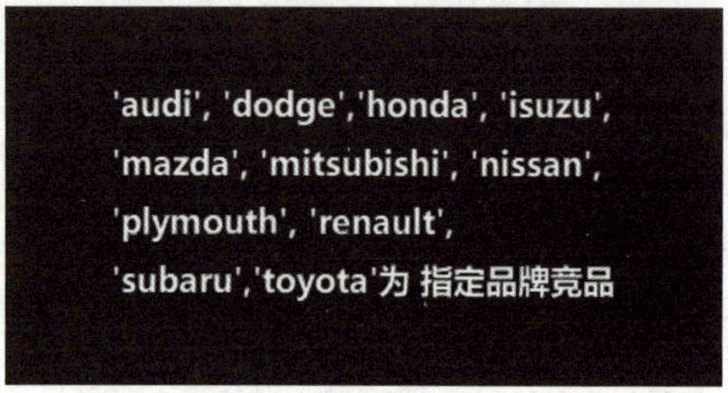

'audi', 'dodge','honda', 'isuzu', 'mazda', 'mitsubishi', 'nissan', 'plymouth', 'renault', 'subaru','toyota'为 指定品牌竞品

图9.6　聚类分析结果

单 元 小 结

　　文本聚类就是将文本视为一个样本，在其上进行聚类操作。聚类对象不是直接的文本本身，而是文本提取出的特征，因此如何提取特征是非常重要的一步。

　　本单元先使同学们对聚类和文本聚类的概念与联系有了一定认知，然后详细介绍了文档特征提取过程中经典的词袋模型的原理和其评价指标。接着引入了 k 均值算法和重复二分聚类算法，并给出了算法的代码实现，使同学们能深入理解其原理并复现。最后设置了文本聚类实战应用，让同学们对文本聚类有更直观的理解。

　　下一单元将在此基础上学习文本分类任务。

习　　题

1．什么是文本聚类？
2．如何实现食品安全评论聚类？

单元10 文本分类

在上一单元中,我们介绍了文本聚类的概念和算法。文本聚类属于粗分类,要想对文本进行精细化分类,需要实现文本分类操作。

本单元将研究文本自动分类问题。在正式介绍文本分类概念之前,需要明确一个概念,那就是监督学习和无监督学习。它们最大的不同是输入的数据集是否有标注。上一单元的文本聚类算法使用的就是无标注数据集,所以它属于无监督学习。在某些场景下,需将文档分类到特定的类别,如过滤垃圾邮件。这种任务就需要一个标注好是否为垃圾邮件的数据集,然后训练分类器学习,最后使用训练好的分类器对测试文档进行预测。这就是典型的文本分类的流程,属于监督学习的范畴。本单元的知识导图如图10.1所示。

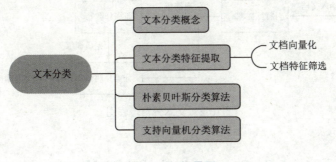

图10.1 知识导图

课程安排

课程任务	课程目标	安排学时
了解文本分类概念	熟悉文本分类概念	1
了解文本分类特征提取	了解文档向量化方法及文档特征筛选方法	1
掌握朴素贝叶斯分类方法	了解朴素贝叶斯分类方法原理,掌握朴素贝叶斯分类方法	1
掌握支持向量机分类方法	了解支持向量机分类方法原理,掌握朴素贝叶斯分类方法	1
掌握文本情感分析方法	了解情感分析概念,掌握情感分析代码实现	1
实战应用	通过该实战应用引导同学们了解一些业务背景,解决实际问题,进行自我练习、自我提高	2

10.1 文本分类概述

首先，给文本分类（Text Classification）一个定义：文本分类是在已知预定义类别的前提下，根据文本的特征、内容或属性，将给定文本与一个或多个类别相关联的过程。比如，判断文档作者在撰写时的情感是"正面"还是"负面"就是一个经典的文本分类任务。除此之外，社交媒体自动标签推荐、对话中的用户意图识别等都属于文本分类任务。

而文本的类别（Class）有时也称标签（Label）。在文本分类任务中，所有的标签都是预定义好的，也就是说模型最终只能在这些标签中选一个作为结果，而不能自己创造新标签。比如，对报纸上的文章主题进行分类，预定义了"金融""社会""政治""体育"四个标签，模型的输出的类别只能四选一，无论如何也不会输出"交通"。但某些特殊的分类模型可能会输出一个特殊的"Unknow"标签，意思是文档不属于任何一种预定义的主题类别，但也不知道其具体类别，所以输出"Unknow"。

一个文本分类的训练和预测的过程可以简略地用图10.2表示。首先对文档语料库进行特征提取，然后将提取到的文档特征按一定比例划分为训练集和测试集，使用测试集的特征与标签对分类器进行训练，再将测试集中的特征输入完成训练的分类器，输出类别作为预测。可以将测试集的预测标签和真实标签进行比较，从而对分类器的性能作出评估。

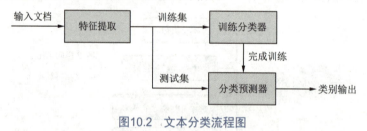

图10.2 文本分类流程图

10.2 文本分类特征提取

10.2.1 文档向量化

与文本聚类一样，若想让机器来自动对文档进行分类，就必须先将文本文档转换为向量表示。一种好的文档向量化的表示方法一方面要能真实地反映文本内容，另一方面要对不同的文本有区分度。

可以沿用9.2.1节中使用的词袋模型来进行文档向量化，在某些教材中词袋模型也被称为向量空间模型（VSM）。虽然它原理简单且不考虑词序，但当搭配TF-IDF或词向量等指标一起使用时，也能有相当好的效果。当某些必须考虑词序的特殊场景下，可以选择使用循环神经网络（RNN）搭配词向量进行建模。RNN有关内容不在教材大纲范围内，同学们若感兴趣可自行了解。

与文本聚类不同的是，文本分类任务中提取到的文档特征有相当一部分可能对分类决策帮助不大，甚至有反作用。如英文中的一些介词或者代词，中文里没有实际意义的虚词等。为了消除这些词的影响，让模型专注于提取有意义的信息，需要借助特征筛选来实现。

10.2.2 文档特征筛选

最简单的文档特征筛选方法是频率法，即人为设定一个上界和下界，删除文档中所有高于上界和低于下界的词。这种做法的动机假设是出现频率过低的词没有代表性，而出现频率过高的词没有区分度。该方法在文档向量化之前使用，可以降低向量计算的复杂度，并可能过滤掉一部分高频和低频噪声提高分类准确率，因为其动机假设还是有一定合理性的。

下面介绍一种功能更强大的特征筛选方法：互信息法。其基本思想是若某特征 f_i 与类别 c_j 的贡献程度越大，那么它们的互信息统计量就越大。假设 N 是数据集中文档的总数，特征全集是 f，类别全集是 C。用 A 表示 C_j 类中包含特征 f_i 的文档数，B 表示除 C_j 之外其他类包含特征 f_i 的文档数，C 表示 C_j 类中不包含特征 f_i 的文档数，D 表示除 C_j 之外其他类不包含特征 f_i 的文档数。$A+B+C+D=N$，具体情况可以见表10.1。

表10.1 特征与类关系

特征	类别	
	C_j	$C-C_j$
f_i	A	B
$f-f_i$	C	D

特征 f_i 与类别 C_j 的互信息可由下列公式计算：

$$I(f_i,C_j) = \log \frac{P(f_i,C_j)}{P(f_i)P(C_j)}$$

$$= \frac{P(f_i|C_j)}{P(f_i)}$$

$$\approx \log \frac{A \cdot N}{(A+C) \cdot (A+B)}$$

如果特征 f_i 与类别 C_j 没有关系，则 $I(f_i,C_j)=0$。在实际应用中判断某特征 f_i 要不要保留时，可以先计算其和所有类别的互信息值然后求平均值，再把这个平均值和预先设好的阈值比较大小。若大于阈值则保留，若小于阈值则丢弃。

10.3 朴素贝叶斯分类算法

朴素贝叶斯分类算法是经典的有监督文本分类算法之一，其原理和代码实现都比较简洁，是初学者必须了解的算法。虽然其名为"朴素"，但在垃圾邮件过滤、情感分析等应用上仍有不俗的表现。朴素贝叶斯分类算法的基本原理是利用贝叶斯定理将特征和类别的联合概率转化为条件概率，再假设各特征之间互相独立来简化条件概率的计算。根据贝叶斯定理，文档 d 属于类别 C_i 的概率是

$$P(C_i|d) = \frac{P(d|C_i) \cdot P(C_i)}{P(d)}$$

由于只需关注上式的相对大小，而 $P(d)$ 是一个和类别无关的常数，所以只需计算分子即可。其中 $P(C_i)$ 这一项很好计算，只需要求每个类别下有多少文档，再将其除以文档总数即可，即

$$P(C_i) = \frac{\text{count}(C_i)}{N}$$

而 $P(d|C_i)$ 这一项可以拆分成已知 C_i 的情况下,d 中每个词 X 出现的联合概率分布,即

$$P(d|C_i) = P(X_1 = x_1, X_2 = x_2, \cdots, X_n = x_n | C_i), d = \{x_1, x_2, \cdots, x_n\}$$

为了简化计算,朴素贝叶斯算法假设所有的特征之间是相互独立的,上式可以转换为每个词出现的条件概率之积,然后再用极大似然法进行近似估计,即

$$P(X_1 = x_1, X_2 = x_2, \cdots, X_n = x_n | C_i) = \prod_{j=1}^{n} P(X_j = x_j | C_i)$$

$$= \prod_{j=1}^{n} \frac{\text{count}(X_j = x_j, C_i)}{\text{count}(C_i)}$$

其中 $\text{count}(C_i)$ 是该类别文档总数;$\text{count}(X_j=x_j, C_i)$ 是 C_i 类别的文章中单词 x_j 出现的次数。至此,算出了文档 d 属于某一类别 C_i 的概率。接下去只需同理计算出所有类别的概率然后把文档分进概率最大的类别即可。

10.4 支持向量机分类算法

支持向量机分类算法(Support Vector Machine,SVM)主要用来解决二分类问题。它的基本思想是在文档向量空间中构造一个最优的决策超平面 A,使其离正负样本的距离最大,从而能最佳地分隔正负类别中的数据点,如图 10.3 所示。训练 SVM 的目标就是找到这个最优的决策边界。

SVM 决策超平面的求解需要高等数学的知识,本书在此不做推导。需要知道 SVM 是一个线性分类器,它的决策边界是一个超平面,只适用于线性可分数据集。而图 10.4 所示的非线性数据,就无法简单地用 SVM 求解。

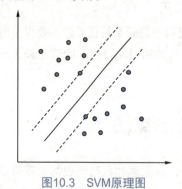

图10.3　SVM原理图　　　　　　图10.4　线性分类器和非线性数据

但如果已知数据在特征空间的分布,就可以构造合适的核函数(Kernel Function)来将特征映射到更高维的空间。让在低维空间线性不可分的数据在高维空间线性可分,就可以使用 SVM 进行求解了。如图 10.5 所示,二维空间中用一个曲面才能分隔的数据,在合理映射到三维空间后用一个平面就可分隔。

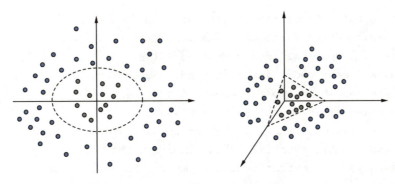

图10.5　二维空间的特征映射

在文本分类中，SVM 算法一般只用于垃圾邮件分类或正负情感分类这种简单的二分类问题；在有需要时，SVM 也可以拓展到多分类的情况，但其不在本书讨论范围内，有兴趣的同学可以自行查阅资料。SVM 的手动实现较为复杂，不过与其相对的是，几乎在所有情况下，SVM 的效果都比朴素贝叶斯更好。包括 HanLP 在内的众多库中均有该分类器，初学者在使用时直接调用 API 使用即可。

10.5　常用文本分类方法

文本分类是自然语言处理中常见的一类任务，为了方便同学们在以后的工作学习中使用，本单元节对常用的文本分类方法进行了整理，并以源码的方式提供给同学们。

1. 数据读取

```python
# 导入必要的库
import os
import shutil
import zipfile
import jieba
import time
import warnings
import xgboost
import lightgbm
import numpy as np
import pandas as pd
from keras import models
from keras import layers
from keras.utils.np_utils import to_categorical
from keras.preprocessing.text import Tokenizer
from sklearn import svm
from sklearn import metrics
from sklearn.neural_network import MLPClassifier
from sklearn.tree import DecisionTreeClassifier
from sklearn.neighbors import KNeighborsClassifier
```

```python
from sklearn.naive_bayes import BernoulliNB
from sklearn.naive_bayes import GaussianNB
from sklearn.naive_bayes import MultinomialNB
from sklearn.linear_model import LogisticRegression
from sklearn.ensemble import RandomForestClassifier
from sklearn.ensemble import AdaBoostClassifier
from sklearn.preprocessing import LabelEncoder
from sklearn.feature_extraction.text import CountVectorizer
from sklearn.feature_extraction.text import TfidfVectorizer

# 路径
filepath = '/原始数据集压缩包路径'
savepath = '/处理后的文件存储路径'
delectpath = '/待处理的文件夹路径'

# ------------------------------------------
# 检索并删除文件夹
if os.path.exists(delectpath):
    print('\n存在该文件夹,正在进行删除,防止解压重命名失败......\n')
    shutil.rmtree(delectpath)
else:
    print('\n不存在该文件夹,请放心处理......\n')

# ------------------------------------------
# 解压并处理中文名字乱码的问题
z = zipfile.ZipFile(filepath, 'r')
for file in z.namelist():
    # 中文乱码需处理
    filename = file.encode('cp437').decode('gbk')   # 先使用 cp437 编码,然后再使用 gbk 解码
    z.extract(file, savepath)                        # 解压 Zip 文件
    # 解决乱码问题
    os.chdir(savepath)                               # 切换到目标目录
    os.rename(file, filename)                        # 将乱码重命名文件

# 定义数据读取函数
def read_text(path, text_list):
    '''
    path: 必选参数,文件夹路径
    text_list: 必选参数,文件夹 path 下的所有 .txt 文件名列表
    return: 返回值
        features 文本(特征)数据,以列表形式返回
        labels 分类标签,以列表形式返回
    '''
```

```python
    features, labels = [], []
    for text in text_list:
        if text.split('.')[-1] == 'txt':
            try:
                with open(path + text, encoding='gbk') as fp:
                    features.append(fp.read())                  # 特征
                    labels.append(path.split('/')[-2])          # 标签
            except Exception as erro:
                print('\n>>> 发现错误,正在输出错误信息...\n', erro)

    return features, labels

def merge_text(train_or_test, label_name):
    '''
    train_or_test: 必选参数,train 训练数据集或 test 测试数据集
    label_name: 必选参数,分类标签的名字
    return: 返回值
        merge_features  合并好的所有特征数据,以列表形式返回
        merge_labels    合并好的所有分类标签数据,以列表形式返回
    '''

    print('\n>>> 文本读取和合并程序已经启动,请稍候...')

    merge_features, merge_labels = [], []         # 函数全局变量
    for name in label_name:
        path = '/home/kesci/work/xiaozhi/text_classification/'+ train_or_test +'/'+ name +'/'
        text_list = os.listdir(path)
        features, labels = read_text(path=path, text_list=text_list)   # 调用函数
        merge_features += features                # 特征
        merge_labels   += labels                  # 标签

    # 可以自定义添加信息
    print('\n>>> 你正在处理的数据类型是...\n', train_or_test)
    print('\n>>>[', train_or_test ,'] 数据具体情况如下...')
    print(' 样本数量 \t', len(merge_features), '\t 类别名称 \t', set(merge_labels))
    print('\n>>> 文本读取和合并工作已经处理完毕...\n')

    return merge_features, merge_labels

# 获取训练集
train_or_test = 'train'
label_name = ['类别名称']
```

```
X_train, y_train = merge_text(train_or_test, label_name)

# 获取测试集
train_or_test = 'test'
label_name = ['类别名称']
X_test, y_test = merge_text(train_or_test, label_name)
```

2. 分词处理

```
# 训练集
X_train_word = [jieba.cut(words) for words in X_train]
X_train_cut = [' '.join(word) for word in X_train_word]

# 测试集
X_test_word = [jieba.cut(words) for words in X_test]
X_test_cut  = [' '.join(word) for word in X_test_word]
```

3. 停词处理

```
# 加载停止词语料
stoplist = [word.strip() for word in open('停词表', \ encoding='utf-8').readlines()]
```

4. 文本标签

```
le = LabelEncoder()
y_train_le = le.fit_transform(y_train)
y_test_le  = le.fit_transform(y_test)

count = CountVectorizer(stop_words=stoplist)
count.fit(list(X_train_cut) + list(X_test_cut))
X_train_count = count.transform(X_train_cut)
X_test_count  = count.transform(X_test_cut)

X_train_count = X_train_count.toarray()
X_test_count  = X_test_count.toarray()

print(X_train_count.shape, X_test_count.shape)
```

5. 算法模型

```
def get_text_classification(estimator, X, y, X_test, y_test):
    '''
    estimator：分类器，必选参数
            X：特征训练数据，必选参数
            y：标签训练数据，必选参数
       X_test：特征测试数据，必选参数
        y_tes：标签测试数据，必选参数
```

```
            return: 返回值
                y_pred_model: 预测值
                    classifier: 分类器名字
                        score: 准确率
                            t: 消耗的时间
                        matrix: 混淆矩阵
                        report: 分类评价函数
    '''
    start = time.time()

    model = estimator
    model.fit(X, y)

    y_pred_model = model.predict(X_test)

    score = metrics.accuracy_score(y_test, y_pred_model)
    matrix = metrics.confusion_matrix(y_test, y_pred_model)
    report = metrics.classification_report(y_test, y_pred_model)

    end = time.time()
    t = end - start
    print('\n>>>算法消耗时间为：', t, '秒\n')
    classifier = str(model).split('(')[0]

    return y_pred_model, classifier, score, round(t, 2), matrix, report
```

(1) K-近邻。

```
knc = KNeighborsClassifier()

result = get_text_classification(knc, X_train_count, y_train_le, X_test_count, y_test_le)
estimator_list.append(result[1]), score_list.append(result[2]), time_list.append(result[3])
```

(2) 多层感知机。

```
mlpc = MLPClassifier()

result = get_text_classification(mlpc, X_train_count, y_train_le, X_test_count, y_test_le)
estimator_list.append(result[1]), score_list.append(result[2]), time_list.append(result[3])
```

（3）高斯。

```
gnb = GaussianNB()

result = get_text_classification(gnb, X_train_count, y_train_le, X_test_count, y_test_le)
    estimator_list.append(result[1]), score_list.append(result[2]), time_list.append(result[3])
```

（4）朴素贝叶斯。

```
mnb = MultinomialNB()

result = get_text_classification(mnb, X_train_count, y_train_le, X_test_count, y_test_le)
    estimator_list.append(result[1]), score_list.append(result[2]), time_list.append(result[3])
```

（5）随机森林。

```
rfc = RandomForestClassifier()

result = get_text_classification(rfc, X_train_count, y_train_le, X_test_count, y_test_le)
    estimator_list.append(result[1]), score_list.append(result[2]), time_list.append(result[3])
```

（6）支持向量回归。

```
svc = svm.SVC()

result = get_text_classification(svc, X_train_count, y_train_le, X_test_count, y_test_le)
    estimator_list.append(result[1]), score_list.append(result[2]), time_list.append(result[3])
```

（7）xgboost。

```
xgb = xgboost.XGBClassifier()

result = get_text_classification(xgb, X_train_count, y_train_le, X_test_count, y_test_le)
    estimator_list.append(result[1]), score_list.append(result[2]), time_list.append(result[3])
```

（8）前馈神经网络。

```
start = time.time()
# np.random.seed(0)                          # 设置随机数种子
feature_num = X_train_count.shape[1]         # 设置所希望的特征数量
```

```python
# 独热编码目标向量来创建目标矩阵
y_train_cate = to_categorical(y_train_le)
y_test_cate = to_categorical(y_test_le)
print(y_train_cate)

# 1. 创建神经网络
network = models.Sequential()

# 2. 添加神经连接层
# 第一层必须有并且一定是输入层，必选
network.add(layers.Dense(     # 添加带有 relu 激活函数的全连接层
                    units=128,
                    activation='relu',
                    input_shape=(feature_num, )
                    ))

# 介于第一层和最后一层之间的称为隐藏层，可选
network.add(layers.Dense(     # 添加带有 relu 激活函数的全连接层
                    units=128,
                    activation='relu'
                    ))
network.add(layers.Dropout(0.8))
# 最后一层必须有并且一定是输出层，必选
network.add(layers.Dense(     # 添加带有 softmax 激活函数的全连接层
                    units=4,
                    activation='sigmoid'
                    ))
# 3. 编译神经网络
network.compile(loss='categorical_crossentropy', # 分类交叉熵损失函数
            optimizer='rmsprop',
            metrics=['accuracy']                 # 准确率度量
            )

# 4. 开始训练神经网络
history = network.fit(X_train_count,             # 训练集特征
            y_train_cate,                        # 训练集标签
            epochs=20,                           # 迭代次数
            batch_size=300,                      # 每个批量的观测数，可做优化
            validation_data=(X_test_count, y_test_cate)   # 验证测试集数据
            )
network.summary()
```

```python
# 5. 模型预测
y_pred_keras = network.predict(X_test_count)

# 6. 性能评估
print('>>> 多分类前馈神经网络性能评估如下...\n')
score = network.evaluate(X_test_count,
                        y_test_cate,
                        batch_size=32)
print('\n>>> 评分 \n', score)
end = time.time()

estimator_list.append('前馈网络')
score_list.append(score[1])
time_list.append(round(end-start, 2))

# 损失函数情况
import matplotlib.pyplot as plt
%matplotlib inline

train_loss = history.history["loss"]
valid_loss = history.history["val_loss"]
epochs = [i for i in range(len(train_loss))]
plt.plot(epochs, train_loss,linewidth=3.0)
plt.plot(epochs, valid_loss,linewidth=3.0)

# 准确率情况
train_loss = history.history["acc"]
valid_loss = history.history["val_acc"]
epochs = [i for i in range(len(train_loss))]
plt.plot(epochs, train_loss,linewidth=3.0)
plt.plot(epochs, valid_loss,linewidth=3.0)

# 7. 保存/加载模型

# 保存
network.save('/home/kesci/work/xiaozhi/my_network_model.h5')

# 加载和使用
my_load_model = models.load_model('/home/kesci/work/xiaozhi/my_network_model.h5')
my_load_model.predict(X_test_count)[:20]
```

（9）LSTM。

```python
# 设置所希望的特征数
feature_num = X_train_count.shape[1]
```

```
# 使用独热编码目标向量对标签进行处理
y_train_cate = to_categorical(y_train_le)
y_test_cate = to_categorical(y_test_le)

# 1. 创建神经网络
lstm_network = models.Sequential()

# 2. 添加神经层
lstm_network.add(layers.Embedding(input_dim=feature_num, output_dim=4))

lstm_network.add(layers.LSTM(units=128))# 添加 128 个单元的 LSTM 神经层

lstm_network.add(layers.Dense(units=4,activation='sigmoid'))
# 添加 sigmoid 分类激活函数的全连接层

# 3. 编译神经网络
lstm_network.compile(loss='binary_crossentropy',optimizer='Adam',metrics=['accuracy'])

# 4. 开始训练模型
lstm_network.fit(X_train_count, y_train_cate, epochs=5, batch_size=128,validation_data=(X_test_count, y_test_cate) )
```

10.6 情感分析

情感分类指根据文本所表达的含义来推测文本作者在撰写时的情感。一般可笼统地划分为正面情感和负面情感；有的数据集也会有多种情感类型，比如电影的评分为 5 分制，就会有 5 个类，但其本质上也是正负情感的细粒度表示。情感分析可以针对词语、句子、段落、文档等任何层级上进行，但不论是什么层级的语料，只要是有标注的数据，就可以将其转换为向量特征，然后当作普通文本分类的问题来解决。

下面以一个包含酒店、电脑、书籍评论的数据集为例，进行情感分析的代码实践。使用 ChnSentiCorp 情感分析数据集进行实验，它包含酒店、电脑及书籍三个行业的评论和对应的情感标签（正面、负面两类）。这是一个句子级的数据集，正负标签比例也较为均衡。这里只以酒店业的评论为例演示。

1. 引入文本分类中要用到的所有库文件

```
from pyhanlp import *
import zipfile
import os
from pyhanlp.static import download
```

2. 下载并解压数据集

```
data_url = "http://file.hankcs.com/corpus/ChnSentiCorp.zip"
root_path = os.path.abspath(os.curdir)
dest_path = os.path.join(root_path, "ChnSentiCorp.zip")
# 把数据集文件下载到当前目录下
download(data_url, dest_path)
# 解压数据集
with zipfile.ZipFile(dest_path, "r") as archive:
    archive.extractall(root_path)
```

解压的数据集就保存在当前目录的 "ChnSentiCorp 情感分析酒店评论" 文件夹内，其包含 "正面" 和 "负面" 两个子文件夹，每个文件夹中各有 2 000 个句子。

正面样例示例：

- 商务大床房，房间很大，床有 2 m 宽，整体感觉经济实惠，不错！
- 周围都在拆迁，环境交通很差，地铁很方便，但是房间很好。
- 地点很方便，房间很舒服，服务也很好，就是价格不便宜啊！

负面样例示例：

- 旅游团住这里的很多，前台服务冷淡。两个人住标准间，只给一张房卡，还很挑衅地看我。气得没心情。
- 服务太不好了，晚上房间的蚊子太多了，人性的关怀不好。
- 卫生间特别小。内部设施不人性化，尤其烧热水的设施明显接触不良,且后面的插线板在下面, 不方便使用。

3. 初始化模型，这里可选朴素贝叶斯或者 SVM，同学们也可以自主尝试其他分类模型

```
# 加载朴素贝叶斯模型参数
NaiveBayesClassifier = JClass('com.hankcs.hanlp.classification.classifiers.NaiveBayesClassifier')

# 重新校准数据集文件
chn_senti_corp = dest_path[:-len('ChnSentiCorp .zip')]+"ChnSentiCorp 情感分析酒店评论 "
```

4. 初始化并训练分类器

```
# 初始化朴素贝叶斯分类器
classifier = NaiveBayesClassifier()
# 训练模型
# classifier 也指出保存模型参数，下次可直接加载参数而无须重复训练
classifier.train(chn_senti_corp)
```

5. 在 predict() 函数中定义输出格式

```
def predict(classifier, text):
    print("《%s》 情感极性是 【%s】" % (text, classifier.classify(text)))
```

6. 自定义一些文本样例并用 predict 函数测试结果

测试结果示意图如图 10.6 所示。

```
>>> predict(classifier,"房间好小，而且地面很脏，真是大失所望!")
《房间好小，而且地面很脏，真是大失所望!》 情感极性是 【负面】
>>> predict(classifier,"服务一流，早餐营养美味，下次还订这家店。")
《服务一流，早餐营养美味，下次还订这家店。》 情感极性是 【正面】
>>> predict(classifier,"高铁二等座的座椅比飞机经济舱舒适很多，好评!")
《高铁二等座的座椅比飞机经济舱舒适很多，好评!》 情感极性是 【正面】
```

图10.6　测试结果示意图

从测试结果可见，我们训练的朴素贝叶斯分类器能比较轻易地分辨酒店业的评论情感的，而且"高铁二等座的座椅比飞机经济舱舒适很多，好评!"属于交通业的评论，我们的模型成功辨别出了它的情感极性，可见该模型具有一定泛化性。

实战应用：新闻标题分类

应用背景：新闻，是人们获取信息、了解时事热点的重要途径。近年来，新闻行业数字化发展迅猛，新闻网络平台的普及，极大地满足了人们"足不出户而知天下事"的心愿。人民网率先建立了以新闻为主的大型网上信息交互平台，是国际互联网上大型综合性网络媒体之一。网络平台上新闻报道、新闻评论、网友发声等文本数据快速增加。将这些文本数据正确归类，可以更好地组织、利用这些信息，因此快速、准确地完成新闻分类任务具有重要的意义。而面对规模巨大且不断增长的文本信息，依靠人工将海量的文本信息分类是不现实的。

自动文本分类技术将人类从烦琐的手工分类中解放出来，使分类任务变得更为高效，为进一步的数据挖掘和分析奠定了基础。公司信息部门希望借助自然语言处理手段实现新闻标题自动分类，以提高工作效率。

1. 数据集

数据集选择 THUCNews，内容是 10 类新闻文本标题的中文分类问题（10 分类），每类新闻标题数据量相等。

2. 数据处理

（1）导入库。

```
import pandas as pd
import numpy as np
import json, time
from tqdm import tqdm
from sklearn.metrics import accuracy_score, classification_report
import torch
import torch.nn as nn
import torch.nn.functional as F
import torch.optim as optim
from torch.utils.data import TensorDataset, DataLoader, RandomSampler, SequentialSampler
```

```python
from transformers import BertModel, BertConfig, BertTokenizer, AdamW, get_cosine_schedule_with_warmup
import warnings
warnings.filterwarnings('ignore')


bert_path = "bert_model/"          # 该文件夹下存放三个文件('vocab.txt', 'pytorch_model.bin', 'config.json')
tokenizer = BertTokenizer.from_pretrained(bert_path)    # 初始化分词器
```

（2）数据预处理。

```python
input_ids, input_masks, input_types,  = [], [], []
# input char ids, segment type ids,  attention mask
labels = []                        # 标签
maxlen = 30                        # 取30即可覆盖99%

with open("news_title_dataset.csv", encoding='utf-8') as f:
    for i, line in tqdm(enumerate(f)):
        title, y = line.strip().split('\t')
        
        # encode_plus 会输出一个字典，分别为'input_ids', 'token_type_ids',
        #'attention_mask' 对应的编码
        # 根据参数会短则补齐，长则切断
        encode_dict = tokenizer.encode_plus(text=title, max_length=maxlen,
                                   padding='max_length', truncation=True)
        
        input_ids.append(encode_dict['input_ids'])
        input_types.append(encode_dict['token_type_ids'])
        input_masks.append(encode_dict['attention_mask'])
        
        labels.append(int(y))
        
input_ids, input_types, input_masks = np.array(input_ids), np.array(input_types), np.array(input_masks)
labels = np.array(labels)
print(input_ids.shape, input_types.shape, input_masks.shape, labels.shape)
```

（3）数据集拆分。

```python
# 随机打乱索引
idxes = np.arange(input_ids.shape[0])
np.random.seed(2019)       # 固定种子
np.random.shuffle(idxes)
print(idxes.shape, idxes[:10])
```

```python
# 8:1:1 划分训练集、验证集、测试集
input_ids_train, input_ids_valid, input_ids_test = input_ids[idxes[:80000]], input_ids[idxes[80000:90000]], input_ids[idxes[90000:]]
input_masks_train, input_masks_valid, input_masks_test = input_masks[idxes[:80000]], input_masks[idxes[80000:90000]], input_masks[idxes[90000:]]
input_types_train, input_types_valid, input_types_test = input_types[idxes[:80000]], input_types[idxes[80000:90000]], input_types[idxes[90000:]]

y_train, y_valid, y_test = labels[idxes[:80000]], labels[idxes[80000:90000]], labels[idxes[90000:]]

print(input_ids_train.shape, y_train.shape, input_ids_valid.shape, y_valid.shape,
      input_ids_test.shape, y_test.shape)
BATCH_SIZE = 64  # 如果会出现 OOM 问题, 则减小它
# 训练集
train_data = TensorDataset(torch.LongTensor(input_ids_train),
                           torch.LongTensor(input_masks_train),
                           torch.LongTensor(input_types_train),
                           torch.LongTensor(y_train))
train_sampler = RandomSampler(train_data)
train_loader = DataLoader(train_data, sampler=train_sampler, batch_size=BATCH_SIZE)

# 验证集
valid_data = TensorDataset(torch.LongTensor(input_ids_valid),
                           torch.LongTensor(input_masks_valid),
                           torch.LongTensor(input_types_valid),
                           torch.LongTensor(y_valid))
valid_sampler = SequentialSampler(valid_data)
valid_loader = DataLoader(valid_data, sampler=valid_sampler, batch_size=BATCH_SIZE)

# 测试集 (是没有标签的)
test_data = TensorDataset(torch.LongTensor(input_ids_test),
                          torch.LongTensor(input_masks_test),
                          torch.LongTensor(input_types_test))
test_sampler = SequentialSampler(test_data)
test_loader = DataLoader(test_data, sampler=test_sampler, batch_size=BATCH_SIZE)
```

3. 模型构建

(1) 定义模型。

```python
# 定义 model
class Bert_Model(nn.Module):
    def __init__(self, bert_path, classes=10):
```

```python
        super(Bert_Model, self).__init__()
        self.config = BertConfig.from_pretrained(bert_path)   # 导入模型超参数
        self.bert = BertModel.from_pretrained(bert_path)      # 加载预训练模型权重
        self.fc = nn.Linear(self.config.hidden_size, classes) # 直接分类


    def forward(self, input_ids, attention_mask=None, token_type_ids=None):
        outputs = self.bert(input_ids, attention_mask, token_type_ids)
        out_pool = outputs[1]          # 池化后的输出 [bs, config.hidden_size]
        logit = self.fc(out_pool)      # [bs, classes]
        return logit
```

如果想要加入 cnn/lstm 等层，可在这里定义。

（2）实例化模型。

```python
def get_parameter_number(model):
    # 打印模型参数量
    total_num = sum(p.numel() for p in model.parameters())
    trainable_num = sum(p.numel() for p in model.parameters() if p.requires_grad)
    return 'Total parameters: {}, Trainable parameters: {}'.format(total_num, trainable_num)


DEVICE = torch.device("cuda" if torch.cuda.is_available() else "cpu")
EPOCHS = 5
model = Bert_Model(bert_path).to(DEVICE)
print(get_parameter_number(model))
```

（3）定义优化器。

```python
optimizer = AdamW(model.parameters(), lr=2e-5, weight_decay=1e-4)  #AdamW 优化器
scheduler = get_cosine_schedule_with_warmup(optimizer, num_warmup_steps=len(train_loader),
         num_training_steps=EPOCHS*len(train_loader))
# 学习率先线性 warmup 一个 epoch，然后 cosine 式下降
```

（4）定义训练函数和测试函数。

```python
# 评估模型性能，在验证集上
def evaluate(model, data_loader, device):
    model.eval()
    val_true, val_pred = [], []
    with torch.no_grad():
        for idx, (ids, att, tpe, y) in (enumerate(data_loader)):
            y_pred = model(ids.to(device), att.to(device), tpe.to(device))
            y_pred = torch.argmax(y_pred, dim=1).detach().cpu().numpy().tolist()
            val_pred.extend(y_pred)
            val_true.extend(y.squeeze().cpu().numpy().tolist())
```

```python
        return accuracy_score(val_true, val_pred)     # 返回accuracy


# 测试集没有标签,需要预测提交
def predict(model, data_loader, device):
    model.eval()
    val_pred = []
    with torch.no_grad():
        for idx, (ids, att, tpe) in tqdm(enumerate(data_loader)):
            y_pred = model(ids.to(device), att.to(device), tpe.to(device))
            y_pred = torch.argmax(y_pred, dim=1).detach().cpu().numpy().tolist()
            val_pred.extend(y_pred)
    return val_pred


def train_and_eval(model, train_loader, valid_loader,
                   optimizer, scheduler, device, epoch):
    best_acc = 0.0
    patience = 0
    criterion = nn.CrossEntropyLoss()
    for i in range(epoch):
        """训练模型"""
        start = time.time()
        model.train()
        print("***** Running training epoch {} *****".format(i+1))
        train_loss_sum = 0.0
        for idx, (ids, att, tpe, y) in enumerate(train_loader):
            ids, att, tpe, y = ids.to(device), att.to(device), tpe.to(device), y.to(device)
            y_pred = model(ids, att, tpe)
            loss = criterion(y_pred, y)
            optimizer.zero_grad()
            loss.backward()
            optimizer.step()
            scheduler.step()                                       # 学习率变化

            train_loss_sum += loss.item()
            if (idx + 1) % (len(train_loader)//5) == 0:            # 只打印五次结果
                print("Epoch {:04d} | Step {:04d}/{:04d} | Loss {:.4f} | Time {:.4f}".format(
                    i+1, idx+1, len(train_loader), train_loss_sum/(idx+1), time.time() - start))
```

```python
                    # print("Learning rate = {}".format(optimizer.state_dict()
['param_groups'][0]['lr']))

        """ 验证模型 """
        model.eval()
        acc = evaluate(model, valid_loader, device)   # 验证模型的性能
        # 保存最优模型
        if acc > best_acc:
            best_acc = acc
            torch.save(model.state_dict(), "best_bert_model.pth")

        print("current acc is {:.4f}, best acc is {:.4f}".format(acc, best_acc))
        print("time costed = {}s \n".format(round(time.time() - start, 5)))
```

4. 模型训练

```python
# 训练和验证评估
train_and_eval(model, train_loader, valid_loader, optimizer, scheduler, DEVICE, EPOCHS)
```

训练结果如图 10.7 所示。

```
***** Running training epoch 1 *****
Epoch 0001 | Step 0250/1250 | Loss 1.6724 | Time 92.7192
Epoch 0001 | Step 0500/1250 | Loss 1.0117 | Time 184.9402
Epoch 0001 | Step 0750/1250 | Loss 0.7427 | Time 276.9679
Epoch 0001 | Step 1000/1250 | Loss 0.6025 | Time 369.3089
Epoch 0001 | Step 1250/1250 | Loss 0.5148 | Time 460.9221
current acc is 0.9607, best acc is 0.9607
time costed = 509.92137s

***** Running training epoch 2 *****
Epoch 0002 | Step 0250/1250 | Loss 0.1184 | Time 92.4572
Epoch 0002 | Step 0500/1250 | Loss 0.1164 | Time 184.3489
Epoch 0002 | Step 0750/1250 | Loss 0.1156 | Time 275.9755
Epoch 0002 | Step 1000/1250 | Loss 0.1142 | Time 368.0724
Epoch 0002 | Step 1250/1250 | Loss 0.1120 | Time 459.7428
current acc is 0.9680, best acc is 0.9680
time costed = 480.61548s
```

图10.7　训练结果

5. 模型测试

```python
# 加载最优权重对测试集测试
model.load_state_dict(torch.load("best_bert_model.pth"))
pred_test = predict(model, test_loader, DEVICE)
print("\n Test Accuracy = {} \n".format(accuracy_score(y_test, pred_test)))
print(classification_report(y_test, pred_test, digits=4))
```

测试结果如图 10.8 所示。由图 10.8 可见，测试准确率为 96.72%。

```
157it [00:19, 7.88it/s]
Test Accuracy = 0.9672
              precision    recall  f1-score   support

           0     0.9947    0.9936    0.9941       938
           1     0.9749    0.9698    0.9723       960
           2     0.9668    0.9788    0.9728      1040
           3     0.9629    0.9590    0.9609      1000
           4     0.9655    0.9636    0.9645       988
           5     0.9952    0.9782    0.9866      1053
           6     0.9432    0.9570    0.9501      1024
           7     0.9405    0.9787    0.9592       985
           8     0.9704    0.9554    0.9629      1031
           9     0.9604    0.9388    0.9495       981

    accuracy                         0.9672     10000
   macro avg     0.9674    0.9673    0.9673     10000
weighted avg     0.9674    0.9672    0.9672     10000
```

图10.8　测试结果

6. 效果展示

程序输出效果如图 10.9 所示。

```
99994    中能电气：先进配电设备制造商    8
99995    以蓝色或红色搭配白色居多（图）    2
```

图10.9　效果展示

经过以上步骤，即可建立起较为完整的文本分类系统。感兴趣的同学可以自己构建数据集，调整参数测试效果。

单 元 小 结

文本分类是指将自然语言或非结构化文本标记为预定义集中的类别的处理。它的应用包括识别产品评论中的积极或者消极情绪、对新闻文章进行分类，以及根据顾客在社交媒体上对产品的交流进行客户细分等。

本单元介绍了文本分类概念，将其和文本聚类概念做了区分，然后介绍了常见的文本向量化、特征筛选等特征提取方法；接着通过对朴素贝叶斯分类算法和支持向量机分类算法原理的阐释，并提供相应的代码实现，使同学们了解这两个经典算法；最后引入了一个情景实战，用实践进一步巩固了理论知识。

下一单元将在文本分类的基础上进行信息挖掘，介绍依存句法分析。

习 题

1．什么是文本分类？常用的文本分类算法有哪些？
2．文本分类有哪些应用领域？以其中一个应用领域为例说明文本分类的具体实现过程。

单元11 依存语法分析

通过分析语言单位中各成分之间的依存关系，可以揭示其句法结构。直观地说，就是对句子中的语法成分进行"主谓宾"和"定状补"的分析，并对各成分之间的相互关系进行分析。分析语法结构，一方面是语言理解本身的需要，语法分析是语言理解的基础；另一方面，语法分析也为其他自然语言处理任务提供了支撑，语法分析是语言理解的基础。

本单元要讲解的语法分析属于文本挖掘的重要内容，介绍了短语结构树和依存语法树两种语法分析的主要模型，解释了依存语法树逐渐成为主流的原因。同学们需要掌握两种语法分析的原理，其中依存语法要掌握词语与词语之间的依存关系，以及如何通过依存关系建立依存语法树；需要重点掌握基于转移的依存语法分析的基本原理以及arc-eager方法的具体思路。在实战模块，本单元还讲解了如何利用pyhanlp库建立依存语法树以及实现意见抽取，同学们可以通过了解其过程实现更广泛的应用。本单元知识导图如图11.1所示。

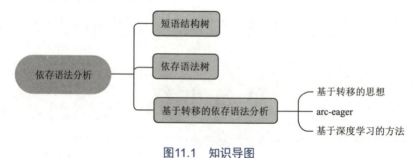

图11.1 知识导图

课程安排

课程任务	课程目标	安排学时
了解短语结构树	了解短语结构树在依存语法分析中的作用	1
了解依存语法树	了解短语结构树在依存语法分析中的作用	1
掌握基于转移的依存语法分析	了解基于转移的依存语法分析原理，并掌握三种主流方法	1
实战应用	通过该实战应用引导同学们了解一些业务背景，解决实际问题，进行自我练习、自我提高	2

11.1 短语结构树

一句话可以根据成分分为主语、谓语、宾语等结构，这个层级之下还有更加细化的句子成分。短语结构语法是将这种句子的成分结构分配给不同的字符串，以使这些字符串能够组成完整的句子的规则系统。它描述了如何自顶向下地生成一个句子，以及如何将句子进行递归分解。短语结构语法是转换生成语法中句法部分的基本部分。

自然语言处理一般用树状图反映句子的句法结构。例如，这句话"我 喜欢 西瓜 果汁"，可以将词语按顺序给词语分配不同的成分结构"专有名词 普通动词 普通名词 普通名词"，将其分解组合可得到图 11.2 所示的短语结构树。

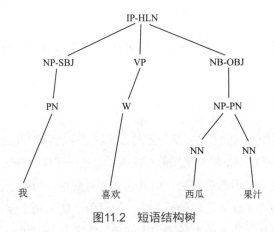

图11.2 短语结构树

其中，IP-HLN 表示单句 - 标题；NP-SBJ 表示名词短语 - 主语；NP-PN 表示名词短语 - 代词。

11.2 依存语法树

短语结构树的规则过于复杂，并且在自然语言处理中的准确度不高，因此大部分研究人员转向研究依存语法树。不同于短语结构树，依存句法树并不关注句子中词语的结构成分以及它们如何组成句子。依存句法树关注的是句子中词语之间的语法联系，并且根据这种联系将句子约束成树状结构。

依存语法理论认为词与词之间存在主从关系，这是一种二元不等价的关系。通常，如果句中的某个词是为了使另一个词意义更完整，则称某个词为从属词，另一个词语称为支配词，两者之间的语法关系称为依存关系。比如，"小幸运"中，形容词"小"与名词"幸运"之间的依存关系如图 11.3 所示，按照一般可视化习惯，箭头指向从属词。

依存句法树就是将句子中存在的所有的依存关系通过树状图的方式呈现出来。根据依存语法树约束性公理，要求除虚根（每句仅一个虚根）外所有单词必须只一次依附于其他单词，并且两个单词之间的词语必须只能依存于两个单词之间的词语（包括这两个单词）。

例如，对于句子"会议 宣布 了 首批 资深 院士 名单"，根据上述规则，可以建立起词语间的依存关系，如图 11.4 所示。

建立起依存语法树，比如句子"会议宣布了首批资深院士名单"的依存句法树（和图 11.4 中的箭头指向相反）如图 11.5 所示。

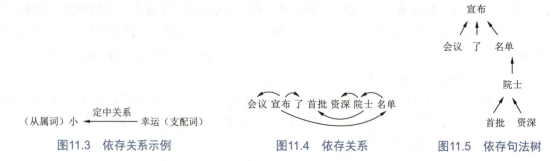

图11.3　依存关系示例　　　　图11.4　依存关系　　　　图11.5　依存句法树

11.3　基于转移的依存语法分析

11.3.1　基于转移的思想

依存语法分析目前有两种主流的方法：基于图的方法和基于转移的方法。基于图的方法准确率相对较高，但其算法复杂度过于复杂，需要全局静态搜索；而基于转移的方法由于采用的局部搜索，容易出现错误传递，准确度不如基于图的方法，但其算法相对简单。与基于图的方法相比，基于转移的方法算法复杂度更低，因此具有更高的分析效率，同时由于能采用更丰富的特征，其分析准确率也与基于图的方法相当，因此受到了越来越多学者的关注，尤其是近年来很多深度学习技术都试图应用于基于转移的方法，因此本书也将重点介绍该类方法。

如果学习模型能够根据句子的某些特征准确地预测如何建立词语之间的依存关系，那么计算机就能够据此拼装出正确的依存句法树。这种拼装动作称为转移（Transition），而这类算法统称为基于转移的依存句法分析。

基于转移的依存句法的主要思路是遍历句子中的每个词，根据词语当前的状态做出决策，例如判断这个词是否与前一个词建立依存关系。这种决策一般是固定的，后面将不会有更改。

该方法通常是从左到右遍历整个句子，对于当前词，通过抽取特征来表示当前状态，然后利用分类器决定要采取的动作。

目前实现依存语法分析主要是 arc-eager 方法与基于深度学习的方法这两个框架。

11.3.2　arc-eager 方法

arc-eager 一般由系统状态的集合、所有可执行的转移动作的集合、初始化函数、终止状态四部分组成。系统状态集合中每个状态都由一个三元组 $<S,I,A>$ 构成，S 是一个栈结构，用于存放已被遍历的词，I 为一个队列结构（缓存区），用于存放句子中还未被分析的词，A 代表已经建立的依存弧集合。该系统包含了四类动作：Left-Arc、Right-Arc、Reduce 和 Shift，见表 11.1。

表11.1 解构组成

动作名称	条件	解释
Shift	队列 β 非空	将队首单词 i 压栈
LeftArc	栈顶单词 i 没有支配词	将栈顶单词 i 的支配词设为队首单词 j，即 i 作为 j 的子节点
RightArc	队首单词 j 没有支配词	将队首单词 j 的支配词设为栈顶单词 i，即 j 作为 i 的子节点
Reduce	栈顶单词 i 已有支配词	将栈顶单词 i 出栈

以"人吃鱼"为例，其转移操作见表 11.2。

表11.2 运行机制

装填编号	σ	转移动作	β	A
0	[]	初始化	[人,吃,鱼,虚根]	{}
1	[人]	Shift	[吃,鱼,虚根]	{}
2	[]	LeftArc（主谓）	[吃,鱼,虚根]	{人←主谓─吃}
3	[吃]	Shift	[鱼,虚根]	{人←主谓─吃}
4	[吃,鱼]	RightArc（动宾）	[虚根]	{人←主谓─吃, 吃─动宾→鱼}
5	[吃]	Reduce	[虚根]	{人←主谓─吃, 吃─动宾→鱼}
6	[]	LeftArc（核红）	[虚根]	{人←主谓─吃, 吃─动宾→鱼, 吃←核心─虚根}

本书中对依存句法树的生成和一系列实验都是借助自然语言处理库 pyhanlp 实现的，具体代码如图 11.6 所示。

```
#模型训练
parser = KBeamArcEagerDependencyParser.train(CTB_TRAIN, CTB_DEV, BROWN_CLUSTER, CTB_MODEL)
#模型运用
print(parser.parse("人吃鱼"))
#模型评估
score = parser.evaluate(CTB_TEST)
```

图11.6 基于转移的依存句法树代码示例

11.3.3 基于深度学习的方法

与传统方法一样，基于深度学习的方法也首先从一个状态中提取一些重要的核心特征，但是与传统方法使用高维、稀疏、离散向量表示（又称 One-hot 表示）这些特征不同，它们使用低维、稠密、连续的分布式向量来表示特征。直觉上，相似的词、词性甚至句法关系等特征可以使用相似的向量表示，从而一定程度上克服数据稀疏问题；理论上，分布式表示是一种降维方法，可有效克服机器学习中的"维数灾难"问题，同时这种分布式表示的表达能力更强，与其维度成指数关系。

深度学习中的非线性激活函数还可以隐含地达到传统方法中特征组合的效果，从而避免烦琐的组合特征设计，最终取得与传统方法相当的准确率。与此同时，该方法由于不需要显式地进行特征组合这一极其耗时的操作，附以预计算等深度学习计算技术，最终还可以极大加快依存句法分析的速度。

下面结合代码介绍基于深度学习的方法。

1. 特征提取

```python
class FeatureExtractor(object):
    def extract_for_current_state(self, sentence, word2idx, pos2idx, dep2idx):
        direct_tokens = self.extract_from_stack_and_buffer(sentence, num_words=3)
        children_tokens = self.extract_children_from_stack(sentence, num_stack_words=2)

        word_features = []
        pos_features = []
        dep_features = []

        # Word features -> 18
        word_features.extend(map(lambda x: x.word, direct_tokens))
        word_features.extend(map(lambda x: x.word, children_tokens))

        # pos features -> 18
        pos_features.extend(map(lambda x: x.pos, direct_tokens))
        pos_features.extend(map(lambda x: x.pos, children_tokens))

        # dep features -> 12 (only children)
        dep_features.extend(map(lambda x: x.dep, children_tokens))

        word_input_ids = [word2idx[word] if word in word2idx else word2idx[UNK_TOKEN.word]
            for word in word_features]
        pos_input_ids = [pos2idx[pos] if pos in pos2idx else pos2idx[UNK_TOKEN.pos]
            for pos in pos_features]
        dep_input_ids = [dep2idx[dep] if dep in dep2idx else dep2idx[UNK_TOKEN.dep]
            for dep in dep_features]

        return [word_input_ids, pos_input_ids, dep_input_ids]  # 48 features

    def extract_from_stack_and_buffer(self, sentence, num_words=3):
        tokens = []

        tokens.extend([NULL_TOKEN for _ in range(num_words - len(sentence.stack))])
        tokens.extend(sentence.stack[-num_words:])

        tokens.extend(sentence.buff[:num_words])
        tokens.extend([NULL_TOKEN for _ in range(num_words - len(sentence.buff))])
        return tokens  # 6 features
```

单元 11 依存语法分析

```python
    def extract_children_from_stack(self, sentence, num_stack_words=2):
        children_tokens = []

        for i in range(num_stack_words):
            if len(sentence.stack) > i:
                lc0 = sentence.get_child_by_index_and_depth(sentence.stack[-i -
1], 0, "left", 1)
                rc0 = sentence.get_child_by_index_and_depth(sentence.stack[-i -
1], 0, "right", 1)

                lc1 = sentence.get_child_by_index_and_depth(sentence.stack[-i -
1], 1, "left", 1) \
                        if lc0 != NULL_TOKEN else NULL_TOKEN
                rc1 = sentence.get_child_by_index_and_depth(sentence.stack[-i -
1], 1, "right", 1) \
                        if rc0 != NULL_TOKEN else NULL_TOKEN

                llc0 = sentence.get_child_by_index_and_depth(sentence.stack[-i -
1], 0, "left", 2) \
                        if lc0 != NULL_TOKEN else NULL_TOKEN
                rrc0 = sentence.get_child_by_index_and_depth(sentence.stack[-i -
1], 0, "right", 2) \
                        if rc0 != NULL_TOKEN else NULL_TOKEN

                children_tokens.extend([lc0, rc0, lc1, rc1, llc0, rrc0])
            else:
                [children_tokens.append(NULL_TOKEN) for _ in range(6)]

        return children_tokens  # 12 features

    def extract_children_from_stack(self, sentence, num_stack_words=2):
        children_tokens = []

        for i in range(num_stack_words):
            if len(sentence.stack) > i:
                lc0 = sentence.get_child_by_index_and_depth(sentence.stack[-i -
1], 0, "left", 1)
                rc0 = sentence.get_child_by_index_and_depth(sentence.stack[-i -
1], 0, "right", 1)

                lc1 = sentence.get_child_by_index_and_depth(sentence.stack[-i -
1], 1, "left", 1) \
                        if lc0 != NULL_TOKEN else NULL_TOKEN
```

```python
                    rc1 = sentence.get_child_by_index_and_depth(sentence.stack[-i -
1], 1, "right", 1)
                        if rc0 != NULL_TOKEN else NULL_TOKEN

                    llc0 = sentence.get_child_by_index_and_depth(sentence.stack[-i -
1], 0, "left", 2)
                        if lc0 != NULL_TOKEN else NULL_TOKEN
                    rrc0 = sentence.get_child_by_index_and_depth(sentence.stack[-i -
1], 0, "right", 2)
                        if rc0 != NULL_TOKEN else NULL_TOKEN

                    children_tokens.extend([lc0, rc0, lc1, rc1, llc0, rrc0])
            else:
                [children_tokens.append(NULL_TOKEN) for _ in range(6)]

    return children_tokens  # 12 features
```

注意：这里实际得到的是 18 个 Token，因此很容易得到 18 个词和词性，以及 12 个 label，最后把它们都变成 id。

2. 数据构建

构建用于训练模型的训练数据。

```python
def create_instances_for_data(self, data, word2idx, pos2idx, dep2idx):
    lables = []
    word_inputs = []
    pos_inputs = []
    dep_inputs = []
    for i, sentence in enumerate(data):
        num_words = len(sentence.tokens)

        for _ in range(num_words * 2):
            word_input, pos_input, dep_input =
            self.extract_for_current_state(sentence, word2idx, pos2idx, dep2idx)
            legal_labels = sentence.get_legal_labels()
            curr_transition = sentence.get_transition_from_current_state()
            if curr_transition is None:
                break
            assert legal_labels[curr_transition] == 1

            # Update left/right children
            if curr_transition != 2:
                sentence.update_child_dependencies(curr_transition)
```

```python
                sentence.update_state_by_transition(curr_transition)
                lables.append(curr_transition)
                word_inputs.append(word_input)
                pos_inputs.append(pos_input)
                dep_inputs.append(dep_input)

        else:
                sentence.reset_to_initial_state()

        # reset stack and buffer to default state
        sentence.reset_to_initial_state()

        targets = np.zeros((len(lables), self.model_config.num_classes), dtype=np.int32)
        targets[np.arange(len(targets)), lables] = 1

        return [word_inputs, pos_inputs, dep_inputs], targets

    def get_legal_labels(self):
        labels = ([1] if len(self.stack) > 2 else [0])
        labels += ([1] if len(self.stack) >= 2 else [0])
        labels += [1] if len(self.buff) > 0 else [0]
        return labels
    def get_transition_from_current_state(self):  # logic to get next transition
        if len(self.stack) < 2:
            return 2  # shift

        stack_token_0 = self.stack[-1]
        stack_token_1 = self.stack[-2]
        if stack_token_1.token_id >= 0 and stack_token_1.head_id == stack_token_0.token_id:
             # left arc
            return 0
        elif stack_token_1.token_id >= -1 and stack_token_0.head_id == stack_token_1.token_id \
            and stack_token_0.token_id not in map(lambda x: x.head_id, self.buff):
            return 1  # right arc
        else:
            return 2 if len(self.buff) != 0 else None

    def get_transition_from_current_state(self):  # logic to get next transition
        if len(self.stack) < 2:
            return 2  # shift
```

```python
        stack_token_0 = self.stack[-1]
        stack_token_1 = self.stack[-2]
        if stack_token_1.token_id >= 0 and stack_token_1.head_id == stack_token_0.token_id:
              # left arc
            return 0
        elif stack_token_1.token_id >= -1 and stack_token_0.head_id == stack_token_1.token_id \
                and stack_token_0.token_id not in map(lambda x: x.head_id, self.buff):
            return 1  # right arc
        else:
            return 2 if len(self.buff) != 0 else None

    def update_child_dependencies(self, curr_transition):
        if curr_transition == 0:
            head = self.stack[-1]
            dependent = self.stack[-2]
        elif curr_transition == 1:
            head = self.stack[-2]
            dependent = self.stack[-1]

        if head.token_id > dependent.token_id:
            head.left_children.append(dependent.token_id)
            head.left_children.sort()
        else:
            head.right_children.append(dependent.token_id)
            head.right_children.sort()

    def update_state_by_transition(self, transition, gold=True):
        if transition is not None:
            if transition == 2:  # shift
                self.stack.append(self.buff[0])
                self.buff = self.buff[1:] if len(self.buff) > 1 else []
            elif transition == 0:  # left arc
                self.dependencies.append((self.stack[-1], self.stack[-2])) if gold else \
                    self.predicted_dependencies.append( (self.stack[-1], self.stack[-2]))
                self.stack = self.stack[:-2] + self.stack[-1:]
            elif transition == 1:  # right arc
                self.dependencies.append((self.stack[-2], self.stack[-1])) if gold else \
                    self.predicted_dependencies.append((self.stack[-2], self.stack[-1]))
                self.stack = self.stack[:-1]
```

3. 网络构建

```python
def build(self):
    self.add_placeholders()
    self.pred = self.add_cube_prediction_op()
    self.loss = self.add_loss_op(self.pred)
    self.accuracy = self.add_accuracy_op(self.pred)
    self.train_op = self.add_training_op(self.loss)

def add_placeholders(self):
    with tf.variable_scope("input_placeholders"):
        self.word_input_placeholder = tf.placeholder(
            shape=[None, self.config.word_features_types],
            dtype=tf.int32, name="batch_word_indices")
        self.pos_input_placeholder = tf.placeholder(
                shape=[None, self.config.pos_features_types],
            dtype=tf.int32, name="batch_pos_indices")
        self.dep_input_placeholder = tf.placeholder(
                shape=[None, self.config.dep_features_types],
            dtype=tf.int32, name="batch_dep_indices")
    with tf.variable_scope("label_placeholders"):
        self.labels_placeholder = tf.placeholder(shape=[None, self.config.num_classes],
            dtype=tf.float32, name="batch_one_hot_targets")
    with tf.variable_scope("regularization"):
        self.dropout_placeholder = tf.placeholder(shape=(), dtype=tf.float32, name="dropout")

def add_cube_prediction_op(self):
    _, word_embeddings, pos_embeddings, dep_embeddings = self.add_embedding()

    with tf.variable_scope("layer_connections"):
        with tf.variable_scope("layer_1"):
            w11 = random_uniform_initializer((self.config.word_features_types *
                self.config.embedding_dim,
                self.config.l1_hidden_size), "w11",
                0.01, trainable=True)
            w12 = random_uniform_initializer((self.config.pos_features_types *
                self.config.embedding_dim,
                self.config.l1_hidden_size), "w12",
                0.01, trainable=True)
            w13 = random_uniform_initializer((self.config.dep_features_types *
                self.config.embedding_dim,
                self.config.l1_hidden_size), "w13",
```

```
                    0.01, trainable=True)
            b1 = random_uniform_initializer((self.config.l1_hidden_size,), "bias1",
                0.01, trainable=True)

            # for visualization
            preactivations = tf.pow(tf.add_n([tf.matmul(word_embeddings, w11),
                tf.matmul(pos_embeddings, w12),
                tf.matmul(dep_embeddings, w13)]) + b1, 3, name="preactivations")

            tf.summary.histogram("preactivations", preactivations)

            h1 = tf.nn.dropout(preactivations,
                keep_prob=self.dropout_placeholder,
            name="output_activations")

        with tf.variable_scope("layer_2"):
            w2 = random_uniform_initializer(
                (self.config.l1_hidden_size, self.config.l2_hidden_size),
                    "w2", 0.01, trainable=True)
            b2 = random_uniform_initializer((self.config.l2_hidden_size,), "bias2",
                0.01, trainable=True)
            h2 = tf.nn.relu(tf.add(tf.matmul(h1, w2), b2), name="activations")

        with tf.variable_scope("layer_3"):
            w3 = random_uniform_initializer(
                    (self.config.l2_hidden_size, self.config.num_classes), "w3",
                0.01, trainable=True)
            b3 = random_uniform_initializer((self.config.num_classes,),
                    "bias3", 0.01, trainable=True)
            with tf.variable_scope("predictions"):
                predictions = tf.add(tf.matmul(h2, w3), b3, name="prediction_logits")

    return predictions

# 除了计算模型输出 pred 的交叉熵 loss 之外，还计算了网络参数的 L2 正则项的 loss
def add_loss_op(self, pred):
    tvars = tf.trainable_variables()
    without_bias_tvars = [tvar for tvar in tvars if 'bias' not in tvar.name]

    with tf.variable_scope("loss"):
        cross_entropy_loss = tf.reduce_mean(tf.nn.softmax_cross_entropy_with_
logits(labels=self.labels_placeholder, logits=pred), name="batch_xentropy_loss")
```

```python
            l2_loss = tf.multiply(self.config.reg_val, self.l2_loss_sum(without_bias_tvars),
                    name="l2_loss")
            loss = tf.add(cross_entropy_loss, l2_loss, name="total_batch_loss")

        tf.summary.scalar("batch_loss", loss)

        return loss

    # 用argmax寻找预测的下标和参考答案labels_placeholder对比得到准确率
    def add_accuracy_op(self, pred):
        with tf.variable_scope("accuracy"):
            accuracy = tf.reduce_mean(tf.cast(tf.equal(tf.argmax(pred, axis=1),
                    tf.argmax(self.labels_placeholder, axis=1)), dtype=tf.float32),
                    name="curr_batch_accuracy")
        return accuracy


    def add_training_op(self, loss):
        with tf.variable_scope("optimizer"):
            optimizer = tf.train.AdamOptimizer(learning_rate=self.config.lr, name="adam_optimizer")
            tvars = tf.trainable_variables()
            grad_tvars = optimizer.compute_gradients(loss, tvars)
            self.write_gradient_summaries(grad_tvars)
            train_op = optimizer.apply_gradients(grad_tvars)

        return train_op
```

4. 模型训练

```python
    def fit(self, sess, saver, config, dataset, train_writer, valid_writer, merged):
        best_valid_UAS = 0
        for epoch in range(config.n_epochs):
            print "Epoch {:} out of {:}".format(epoch + 1, self.config.n_epochs)

            summary, loss = self.run_epoch(sess, config, dataset, train_writer, merged)

            if (epoch + 1) % dataset.model_config.run_valid_after_epochs == 0:
                valid_UAS = self.run_valid_epoch(sess, dataset)
                valid_UAS_summary = tf.summary.scalar("valid_UAS",
                        tf.constant(valid_UAS, dtype=tf.float32))
                valid_writer.add_summary(sess.run(valid_UAS_summary), epoch + 1)
                if valid_UAS > best_valid_UAS:
```

```python
                    best_valid_UAS = valid_UAS
                    if saver:
                        print "New best dev UAS! Saving model.."
                        saver.save(sess, os.path.join(DataConfig.data_dir_path,
                DataConfig.model_dir, DataConfig.model_name))

            # trainable variables summary -> only for training
            if (epoch + 1) % dataset.model_config.write_summary_after_epochs == 0:
                train_writer.add_summary(summary, global_step=epoch + 1)

    def train_on_batch(self, sess, inputs_batch, labels_batch, merged):
        word_inputs_batch, pos_inputs_batch, dep_inputs_batch = inputs_batch
        feed = self.create_feed_dict([word_inputs_batch, pos_inputs_batch, dep_inputs_batch],labels_batch=labels_batch,keep_prob=self.config.keep_prob)
        _, summary, loss = sess.run([self.train_op, merged, self.loss], feed_dict=feed)
        return summary, loss
```

5. 模型测试

```python
    def run_valid_epoch(self, sess, dataset):
        print "Evaluating on dev set",
        self.compute_dependencies(sess, dataset.valid_data, dataset)
        valid_UAS = self.get_UAS(dataset.valid_data)
        print "- dev UAS: {:.2f}".format(valid_UAS * 100.0)
        return valid_UAS

    def compute_dependencies(self, sess, data, dataset):
        sentences = data
        rem_sentences = [sentence for sentence in sentences]
        [sentence.clear_prediction_dependencies() for sentence in sentences]
        [sentence.clear_children_info() for sentence in sentences]

        while len(rem_sentences) != 0:
            curr_batch_size = min(dataset.model_config.batch_size, len(rem_sentences))
            batch_sentences = rem_sentences[:curr_batch_size]

            enable_features = [0 if len(sentence.stack) == 1 and len(sentence.buff) == 0 else 1
                               for sentence in batch_sentences]
            enable_count = np.count_nonzero(enable_features)

            while enable_count > 0:
```

```python
curr_sentences = [sentence for i, sentence in enumerate(batch_sentences) if
    enable_features[i] == 1]


curr_inputs = [
dataset.feature_extractor.extract_for_current_state(sentence, dataset.word2idx,
dataset.pos2idx, dataset.dep2idx) for sentence in curr_sentences]
word_inputs_batch = [curr_inputs[i][0] for i in range(len(curr_inputs))]
pos_inputs_batch = [curr_inputs[i][1] for i in range(len(curr_inputs))]
dep_inputs_batch = [curr_inputs[i][2] for i in range(len(curr_inputs))]

predictions = sess.run(self.pred,
   feed_dict=self.create_feed_dict([word_inputs_batch, pos_inputs_batch,
      dep_inputs_batch]))
legal_labels = np.asarray([sentence.get_legal_labels() for sentence in
curr_sentences],
        dtype=np.float32)
legal_transitions = np.argmax(predictions + 1000 * legal_labels, axis=1)

# update left/right children so can be used for next feature vector
[sentence.update_child_dependencies(transition) for (sentence, transition) in
    zip(curr_sentences, legal_transitions) if transition != 2]

# update state
[sentence.update_state_by_transition(legal_transition, gold=False)
for (sentence, legal_transition) in zip(curr_sentences, legal_transitions)]

enable_features = [0 if len(sentence.stack) == 1 and len(sentence.buff) == 0 else 1
    for sentence in batch_sentences]
enable_count = np.count_nonzero(enable_features)

# Reset stack and buffer
[sentence.reset_to_initial_state() for sentence in batch_sentences]
rem_sentences = rem_sentences[curr_batch_size:]
```

实战应用：基于依存语法树的意见抽取

应用背景：前面的单元中构建了商品评价意见提取模型，基于该模型的产品上线后广受好评。但是，随着用户数量的增加，也暴露出算法中的一些不足，如在某些情形下，无法处理句式灵活的文本，导致提取的评价意见不够准确全面。经理希望项目团队能够优化并解决这一问题，并尽快上线测试。

本实战目的是提取商品评论中的属性和买家评价。

以下述评价为例：
>眼影盘颜色很好看，显色度很好，但容易飞粉，着色度一般。
这时就可以对这句话进行依存句法分析，分析代码如下：

```python
from pyhanlp import *

CoNLLSentence = JClass('com.hankcs.hanlp.corpus.dependency.CoNll.CoNLLSentence')
CoNLLWord = JClass('com.hankcs.hanlp.corpus.dependency.CoNll.CoNLLWord')
IDependencyParser = JClass('com.hankcs.hanlp.dependency.IDependencyParser')
KBeamArcEagerDependencyParser = JClass('com.hankcs.hanlp.dependency.perceptron.parser.KBeamArcEagerDependencyParser')

parser = KBeamArcEagerDependencyParser()
tree = parser.parse("眼影颜色很好看，显色度很好，但容易飞粉，着色度一般。")
print(tree)
def extactOpinion3(tree):
    for word in tree.iterator():
        if word.POSTAG == "NN":

            # 检测名词词语的依存弧是否是"属性关系"，
            # 如果是，则寻找支配词的子节点中的主题词
            # 以该主题词作为名词的意见。
            if word.DEPREL == "nsubj":                  # ①属性

                if tree.findChildren(word.HEAD, "neg").isEmpty():
                    print("%s = %s" % (word.LEMMA, word.HEAD.LEMMA))
                else:
                    print("%s = 不%s" % (word.LEMMA, word.HEAD.LEMMA))
            elif word.DEPREL == "attr":
                top = tree.findChildren(word.HEAD, "top")       # ②主题

                if not top.isEmpty():
                    print("%s = %s" % (word.LEMMA, top.get(0).LEMMA))

print("第三版")
extactOpinion3(tree)
```

单 元 小 结

语法分析是自然语言处理中一个重要的任务，其目标是分析句子的语法结构，并将其表示为容易理解的结构（通常是树状结构）。

本单元介绍了语法分析的两大方法——短语结构分析和依存句法分析，详细阐述了这两种方法

的基础原理，让同学们了解二者的不同侧重。对于基于依存语法的实现本文简述了基于图以及基于转移的两种不同方式，详细介绍了基于转移的方法，让同学们了解其遍历的主要思路以及 arc-eager 方法如何实现依存句法分析。在最后应用实例中介绍了依存语法的应用场景——意见抽取，让同学们了解如何利用 pyhanlp 库建立依存语法树并实现意见抽取。

下一单元将学习自然语言处理中的一项高级任务——机器翻译。

习　题

1. 什么是依存语法分析？举例说明依存语法分析的应用场景。
2. 什么是依存句法树？如何构建依存句法树？

单元12 深度学习与自然语言处理

通过前面单元的学习，相信同学们对于自然语言处理有了一定的整体认识，并能够运用自然语言处理这一工具解决各类问题。在前述单元的学习中侧重于介绍传统自然语言处理方法，本单元将着重介绍运用深度学习方法解决自然语言处理问题。

深度学习(Deep Learning)是机器学习领域的一个新概念，是一种用于建立和模仿人类大脑机制的机器学习领域的新技术，常用于机器学习、自然语言处理等领域。

本单元阐述了从自然语言处理发展到深度学习的过程和其使用过程中所需要用到的概念、函数及工具包，并能够运用深度学习方法解决机器翻译问题，要求同学们了解深度学习和自然语言处理的应用场景及主要任务。本单元知识导如图12.1所示。

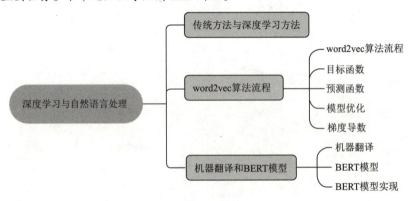

图12.1 知识导图

课程安排

课程任务	课程目标	安排学时
了解深度学习方法	了解深度学习方法和传统自然语言处理方法的区别	1
掌握word2vec方法	通过对word2vec掌握深度学习方法的基本思路	1
掌握BERT模型	掌握BERT模型的原理并能运用BERT模型解决实际问题	1
实战应用	通过该实战应用引导同学们了解一些业务背景，解决实际问题，进行自我练习、自我提高	2

12.1 传统方法与深度学习方法

12.1.1 传统方法

NLP问题的传统或经典的解决方法是几个步骤组成的流程工作，它是一种统计方法。相同的NLP任务若要获得良好的性能，特征提取是最耗时且最关键的步骤。

传统方法的处理流程简单来说就是：特征提取 + 传统模型训练，如图12.2所示。

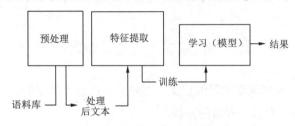

图12.2 传统方法处理流程图

12.1.2 深度学习方法

深度学习是机器学习的一个分支，它更侧重于尝试自动学习合适特征，尝试学习多层次的表征及输出。

深度学习优势总结：手工特征消耗的时间精力较大且不易拓展，相比之下，自动特征学习速度快且便于拓展；深度学习提供了一种通用的学习框架，既可以进行无监督学习，也可以进行监督学习。

深度学习已成功应用于自然语言处理中，并取得了重大进展。与传统方法相比，深度学习的重要特点是用向量表示各级元素，通过传统方法标注深度学习，使用向量表示单词、词语逻辑、表达式和句子，搭建多层神经网络，让机器能进行自主学习。

此外，深度学习就像一个黑盒子，人们把数据输入进去，但却无法得知这些数据在深度学习的过程中经过了哪些处理，仅仅只能看到它最后得出的结果，如图12.3所示。但是，当这个过程中出现问题的时候，人们并不知道该如何去调整。

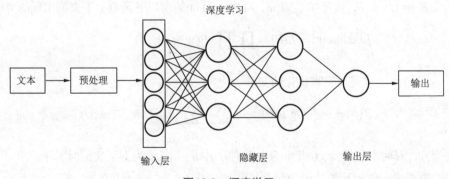

图12.3 深度学习

相比之下，传统的算法就会比较有优势，如图12.4所示，可以通过一些过程的分析，查到这个数据里的一些词的分布，如果这个数据里出现异样，同样可以通过这些去进行分析找出问题，然后调整数据。

图12.4 传统自然语言处理

在处理一些比较经典的任务时，如文本分类、命名实体识别等，传统算法也是能取得不错效果的。在进行方法选择时，需结合多方条件来进行选择，以此来达到更加良好的结果。

12.2　word2vec（词向量）

作为连接传统机器学习与深度学习的桥梁，词向量一直是入门深度学习的第一站。词向量的训练方法有很多种，word2vec 是其中经典的一种。

12.2.1　word2vec 算法流程

从随机词向量开始，反复阅读整个语料库中的每个单词，使用单词向量预测周围的单词。

优化过程：更新向量，以便更好地预测实际周围的单词。该算法学习的单词向量能够很好地捕捉单词相似性和语义，如图 12.5 所示。

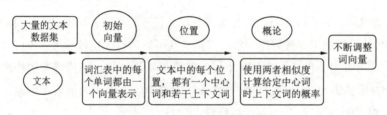

图12.5　word2vec算法流程图

12.2.2　目标函数

对每个位置 $t=1,\cdots,T$，给定中心词 w_j，在大小为 m 的窗口下预测上下文词出现的可能性，即

$$\text{Likelihood} = L(\theta) = \prod_{t=1}^{T} \prod_{\substack{m \leqslant j \leqslant m \\ j \neq 0}} P(w_{t+j} \mid w_t; \theta)$$

θ 是所有被优化的变量

$$J(\theta) = -\frac{1}{T} \log L(\theta) = -\frac{1}{T} \sum_{t=1}^{T} \sum_{\substack{m \leqslant j \leqslant m \\ j \neq 0}} \log P(w_{t+j} \mid w_t; \theta)$$

对于每个 w，有两个向量：v_w（当 w 为中心词时）、u_w（当 w 为上下文词时）。

对于一个中心词 c 和上下文词 o，有

$$P(o \mid c) = \frac{\exp(u_o^{\mathrm{T}} v_c)}{\sum_{w \in V} \exp(u_w^{\mathrm{T}} v_c)}$$

计算 $P(w_{t+j}|w_t)$，在大小为 m 的窗口下预测上下文词出现的可能性。

用 $P(u_{\text{problems}}|v_{\text{into}})$ 代替 $P(\text{problems}|\text{into}; u_{\text{problems}}, v_{\text{into}}, \theta)$，如图 12.6 所示。

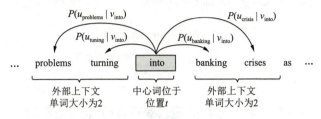

图12.6　计算 $P(w_{t+j}|w_t)$

12.2.3　预测函数

softmax() 函数：将任意值 x_i 映射到概率 p_i 上，即

$$\text{softmax}(x_i) = \frac{\exp(x_i)}{\sum_{j=1}^{n}\exp(x_j)} = p_i$$

max 放大了较大的 x_i 的可能性；soft 仍为较小的 x_i 保留可能性；softmax 常应用于深度学习中。

12.2.4　模型优化

为了训练一个模型，逐步调整参数以使损失最小化，如图 12.7 所示。

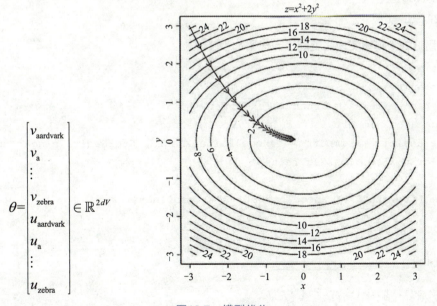

图12.7　模型优化

θ 在一个长向量中表示所有模型参数、d 维向量和 V 个单词，每个单词都有两个向量沿着梯度向下方向优化参数，计算所有向量梯度。

12.2.5 梯度导数

链式法则：定义 $y=f(u)$ 且 $u=g(x)$，即 $y=f(g(x))$

$$\log p(o|c) = \log \frac{\exp(u_o^T v_c)}{\sum_{w=1}^{V} \exp(u_w^T v_c)}$$

12.3 机器翻译和BERT模型

1. 机器翻译

机器翻译是指利用计算机将一种语言转换成另一种语言。在当下经济全球化发展时期，机器翻译在多个方面起到了越来越重要的作用。

机器翻译的方法总的来说分为两种：基于规则、基于语料库，而基于语料库的方法又包含了统计和神经两种。

2. BERT 模型

BERT 模型是一个预训练的语言模型，任何使用者都可以直接继承前人所建立好的模型，继续运用。但可复现性差，只能直接用。相对于 Transformer 这种浅而宽的模型，BERT 这种深而窄的模型预训练的结果会更好。同时，BERT 模型是一个基于 Transformer 的双向编码器，它拥有双向编码能力和特征提取能力。

3. BERT 模型实现

下面使用 BERT 模型实现法英翻译。代码如下：

```
import os
import numpy as np
from transformers import AutoTokenizer
from datasets import load_dataset,load_metric
from transformers import AutoModelForSeq2SeqLM, Seq2SeqTrainingArguments,DataCollatorForSeq2Seq, Seq2SeqTrainer

f=open("fra.txt","r",encoding="utf-8").readlines()
en=[]
fre=[]
data=[]
for l in f:
    line=l.strip().split("\t")
    tmp={}
    tmp["en"]=line[0]
    tmp["fr"]=line[1]
    data.append(tmp)
print(len(data))
```

```python
print(data[0])
170651
{'en':'Go.', 'fr': 'Va!'}

train_size=int(len(data)*0.9)
train_data=data[:train_size]
val_data=data[train_size:]

f=open("train.txt","w" , encoding='utf-8')
for i in train_data:
    f.write(str(i)+"\n")
f.close()

f=open("val.txt","w" , encoding='utf-8')
for i in val_data:
    f.write(str(i)+"\n")
f.close()

# 分词，使用已经训练好的helsinki-NLP/opus-mt-en-ro来做翻译任务
model_checkpoint="Helsinki-NLP/opus-mt-en-ro"
tokenizer=AutoTokenizer.from_pretrained(model_checkpoint,use_fast=True)
raw_datasets=load_dataset("text",data_files={"train":"train.txt","validation":"val.txt"})

max_input_length=64
max_target_length=64
source_lang="en"
target_lang="fr"

def preprocess_function(examples):
    inputs=[eval(ex)[source_lang] for ex in examples["text"]]
    targets=[eval(ex)[target_lang] for ex in examples["text"]]
    model_inputs=tokenizer(inputs,max_length=max_input_length,truncation=True)
    # 为目标语言设置分词器
    with tokenizer.as_target_tokenizer():
        labels=tokenizer(targets,max_length=max_target_length,truncation=True)
    model_inputs["labels"]=labels["input_ids"]
    return model_inputs
# 把预处理函数应用到原始的dataset中
tokenized_datasets=raw_datasets.map(preprocess_function,batched=True)
# 加载预训练模型
model=AutoModelForSeq2SeqLM.from_pretrained(model_checkpoint)
# 设定训练参数
```

```
batch_size=8
args=Seq2SeqTrainingArguments(
    "test-translation",
    evaluation_strategy="epoch",
    learning_rate=2e-5,
    per_device_train_batch_size=batch_size,
    per_device_eval_batch_size=batch_size,
    weight_decay=0.01,
    save_total_limit=3, # 至多保存的模型个数
    num_train_epochs=10,
    predict_with_generate=True,
    fp16=False,
)

# 数据收集器 data collator，告诉 trainer 如何从预处理的输入数据中构造 batch，使用数据处理器 DataCollatorForSeq2Seq
# 将预处理的输入分 batch 再次处理后输入模型
data_collator=DataCollatorForSeq2Seq(tokenizer,model=model)

# 定义评估方法，使用 bleu 指标，利用 metric.compute 计算该指标对模型进行评估
# 定义 postprocess_text 函数做一些数据后处理
metric=load_metric("sacrebleu")
print("successfully import metric")

def postprocess_text(preds, labels):
    preds = [pred.strip() for pred in preds]
    labels = [[label.strip()] for label in labels]
    return preds, labels

def compute_metrics(eval_preds):
    preds, labels = eval_preds
    if isinstance(preds, tuple):
        preds = preds[0]
    decoded_preds = tokenizer.batch_decode(preds, skip_special_tokens=True)

    # 如果 labels=-100，说明这个 label 是无法编码的，应该用 pad_token_id 进行填充
    labels = np.where(labels != -100, labels, tokenizer.pad_token_id)
    decoded_labels = tokenizer.batch_decode(labels, skip_special_tokens=True)

    # 把预测后的输出转成适合 metric.compute 输入的格式
    print("type(decoded_preds)=", type(decoded_preds))
```

```python
        print("type(decoded_labels)=", type(decoded_labels))
        decoded_preds, decoded_labels = postprocess_text(decoded_preds, decoded_labels)

        result = metric.compute(predictions=decoded_preds, references=decoded_labels)
        result = {"bleu": result["score"]}

        prediction_lens = [np.count_nonzero(pred != tokenizer.pad_token_id) for pred in preds]
        result["gen_len"] = np.mean(prediction_lens)
        result = {k: round(v, 4) for k, v in result.items()}
        print("result is as follow=============================")
        print(result)
        return result

# 开始训练
os.environ['CUDA_VISIBLE_DEVICES'] = '0'

trainer=Seq2SeqTrainer(
    model,
    args,
    train_dataset=tokenized_datasets["train"],
    eval_dataset=tokenized_datasets["validation"],
    data_collator=data_collator,
    tokenizer=tokenizer,
    compute_metrics=compute_metrics
)
trainer.train()

type(decoded_preds)=<class 'list'>
type(decoded_labels)=<class 'list'>
result is as follow=======================
{'bleu': 40.4785, 'gen_len': 28.4694}
type(decoded_preds)=<class 'list'>
type(decoded_labels)=<class 'list'>
result is as follow=======================
{'bleu': 44.3371, 'gen_len': 28.34}
TrainOutput(global_step=24000, training_loss=0.3073067795435588, metrics={'train_runtime': 9196.9589, 'train_samples_per_second': 166.995, 'train_steps_per_second': 2.61, 'total_flos': 5799969455210496.0, 'train_loss': 0.3073067795435588, 'epoch': 10.0})
```

实战应用：使用神经网络实现英文-中文翻译

应用背景：机机器翻译（Machine Translation）又称自动翻译，是利用计算机把一种自然语言（源语）转化为另一种自然语言（目标语）的过程，具有重要科学研究价值。同时，机器翻译的实用性也很强，在推动政治、经济和文化交流方面发挥着越来越重要的作用，肩负着架起语言沟通桥梁的重任。

作为一家新媒体技术公司，为了提升用户的使用体验，项目经理希望能够上线翻译功能，辅助客户。请利用所学知识，结合场景需求，构建算法模型。

1. 文本数据处理

（1）数据预处理。

原始数据格式如图 12.8 所示。

```
Hi.       嗨。         CC-BY 2.0 (France) Attribution: tatoeba.org #538123 (CM) & #891077 (Martha)
Hi.       你好。        CC-BY 2.0 (France) Attribution: tatoeba.org #538123 (CM) & #4857568 (musclegirlxyp)
Run.      你用跑的。     CC-BY 2.0 (France) Attribution: tatoeba.org #4008918 (JSakuragi) & #3748344 (eg
Wait!     等等！        CC-BY 2.0 (France) Attribution: tatoeba.org #1744314 (belgavox) & #4970122 (wzhd)
Wait!     等一下！      CC-BY 2.0 (France) Attribution: tatoeba.org #1744314 (belgavox) & #5092613 (mirrorvan)
Hello!    你好。        CC-BY 2.0 (France) Attribution: tatoeba.org #373330 (CK) & #4857568 (musclegirlxyp)
I try.    让我来。      CC-BY 2.0 (France) Attribution: tatoeba.org #20776 (CK) & #5092185 (mirrorvan)
I won!    我赢了。      CC-BY 2.0 (France) Attribution: tatoeba.org #2005192 (CK) & #5102367 (mirrorvan)
Oh no!    不会吧。      CC-BY 2.0 (France) Attribution: tatoeba.org #1299275 (CK) & #5092475 (mirrorvan)
Cheers!   干杯！        CC-BY 2.0 (France) Attribution: tatoeba.org #487006 (human600) & #765577 (Martha)
Got it?   你懂了吗？    CC-BY 2.0 (France) Attribution: tatoeba.org #455353 (FeuDRenais) & #7768205 (jia
He ran.   他跑了。      CC-BY 2.0 (France) Attribution: tatoeba.org #672229 (CK) & #5092389 (mirrorvan)
Hop in.   跳进来。      CC-BY 2.0 (France) Attribution: tatoeba.org #1111548 (Scott) & #5092444 (mirrorvan)
I quit.   我退出。      CC-BY 2.0 (France) Attribution: tatoeba.org #731636 (Eldad) & #5102253 (mirrorvan)
I'm OK.   我沒事。      CC-BY 2.0 (France) Attribution: tatoeba.org #433763 (CK) & #819304 (Martha)
Listen.   听着。        CC-BY 2.0 (France) Attribution: tatoeba.org #1913088 (CK) & #5092137 (mirrorvan)
No way!   不可能！      CC-BY 2.0 (France) Attribution: tatoeba.org #2175 (CS) & #503298 (fucongcong)
Really?   你确定？      CC-BY 2.0 (France) Attribution: tatoeba.org #373216 (kotobaboke) & #4208543 (Ethan_lin)
Try it.   试试吧。      CC-BY 2.0 (France) Attribution: tatoeba.org #4756252 (cairnhead) & #4757237 (ryanwoo)
```

图12.8 原始数据

导入必要的库：

```python
import pandas as pd
import numpy as np
from keras.layers import Input, LSTM, Dense, merge,concatenate
from keras.optimizers import Adam, SGD
from keras.models import Model,load_model
from keras.utils import plot_model
from keras.models import Sequential

# 定义神经网络的参数
NUM_SAMPLES=3000         # 训练样本的大小
batch_size = 64          # 一次训练所选取的样本数
epochs = 100             # 训练轮数
latent_dim = 256         # LSTM 的单元个数
```

读取数据文件：

```
data_path='data/cmn.txt'
df=pd.read_table(data_path,header=None).iloc[:NUM_SAMPLES,0:2]

# 添加标题栏
df.columns=['inputs','targets']

# 每句中文举手加上 '\t' 作为起始标志，句末加上 '\n' 终止标志
df['targets']=df['targets'].apply(lambda x:'\t'+x+'\n')latent_dim = 256
# LSTM 的单元个数
```

读入后的数据格式如图 12.9 所示。

Index	inputs	targets
0	Hi.	嗨。
1	Hi.	你好。
2	Run.	你用跑的。
3	Wait!	等等！
4	Wait!	等一下！
5	Hello!	你好。
6	I try.	让我来。
7	I won!	我赢了。
8	Oh no!	不会吧。
9	Cheers!	干杯！
10	Got it?	你懂了吗？

图12.9　读入后的数据

对数据格式进行转换：

```
# 获取英文、中文各自的列表
input_texts=df.inputs.values.tolist()
target_texts=df.targets.values.tolist()

# 确定中英文各自包含的字符。df.unique() 直接取 sum 可将 unique 数组中的各个句子拼接成一个长句子
input_characters = sorted(list(set(df.inputs.unique().sum())))
target_characters = sorted(list(set(df.targets.unique().sum())))

# 英文字符中不同字符的数量
num_encoder_tokens = len(input_characters)

# 中文字符中不同字符的数量
```

```
num_decoder_tokens = len(target_characters)

# 最大输入长度
INUPT_LENGTH = max([ len(txt) for txt in input_texts])

# 最大输出长度
OUTPUT_LENGTH = max([ len(txt) for txt in target_texts])
```

(2)创建字典。

```
input_token_index = dict( [(char, i)for i, char in enumerate(input_characters)] )
target_token_index = dict( [(char, i) for i, char in enumerate(target_characters)] )

reverse_input_char_index = dict([(i, char) for i, char in enumerate(input_characters)])
reverse_target_char_index = dict([(i, char) for i, char in enumerate(target_characters)])
```

(3)对英文-中文句子进行编码。

```
# 需要把每条语料转换成LSTM需要的三维数据输入[n_samples, timestamp, one-hot feature]到模型中
encoder_input_data =np.zeros((NUM_SAMPLES,INUPT_LENGTH,num_encoder_tokens))
decoder_input_data =np.zeros((NUM_SAMPLES,OUTPUT_LENGTH,num_decoder_tokens))
decoder_target_data = np.zeros((NUM_SAMPLES,OUTPUT_LENGTH,num_decoder_tokens))

for i,(input_text,target_text) in enumerate(zip(input_texts,target_texts)):
    for t,char in enumerate(input_text):
        encoder_input_data[i,t,input_token_index[char]]=1.0

    for t, char in enumerate(target_text):
        decoder_input_data[i,t,target_token_index[char]]=1.0

        if t > 0:
            # decoder_target_data 不包含开始字符,并且比decoder_input_data提前一步
            decoder_target_data[i, t-1, target_token_index[char]] = 1.0
```

2. 模型构建

```
def create_model():
    # 定义编码器的输入
    encoder_inputs=Input(shape=(None,num_encoder_tokens))

    # 返回状态
    encoder=LSTM(latent_dim,return_state=True)

    # 调用编码器,得到编码器的输出,以及状态信息 state_h 和 state_c
    encoder_outputs,state_h,state_c=encoder(encoder_inputs)
```

```python
# 丢弃 encoder_outputs, 保存 state_h, state_c
encoder_state=[state_h,state_c]

# 定义解码器的输入
decoder_inputs=Input(shape=(None,num_decoder_tokens))

# 并且返回其中间状态，中间状态在训练阶段不会用到，但在推理阶段有用
decoder_lstm=LSTM(latent_dim,return_state=True,return_sequences=True)

# 将编码器输出的状态作为初始解码器的初始状态
decoder_outputs,_,_=decoder_lstm(decoder_inputs,initial_state=encoder_state)

# 添加全连接层
decoder_dense=Dense(num_decoder_tokens,activation='softmax')
decoder_outputs=decoder_dense(decoder_outputs)

# 定义整个模型
model=Model([encoder_inputs,decoder_inputs],decoder_outputs)

# 定义 sampling 模型
# 定义 encoder 模型，得到输出 encoder_states
encoder_model=Model(encoder_inputs,encoder_state)

decoder_state_input_h=Input(shape=(latent_dim,))
decoder_state_input_c=Input(shape=(latent_dim,))
decoder_state_inputs=[decoder_state_input_h,decoder_state_input_c]

# 得到解码器的输出以及中间状态
    decoder_outputs,state_h,state_c=decoder_lstm(decoder_inputs,initial_state=decoder_state_inputs)
    decoder_states=[state_h,state_c]
    decoder_outputs=decoder_dense(decoder_outputs)
    decoder_model=Model([decoder_inputs]+decoder_state_inputs,[decoder_outputs]+decoder_states)

    plot_model(model=model,show_shapes=True)
    plot_model(model=encoder_model,show_shapes=True)
    plot_model(model=decoder_model,show_shapes=True)
    return model,encoder_model
```

使用的模型结构如图 12.10 所示。

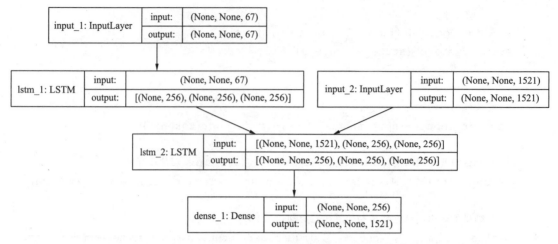

图12.10 模型结构

3. 模型训练

```
def train():
    model,encoder_model,decoder_model=create_model()

    # 编译模型
    model.compile(optimizer='rmsprop',loss='categorical_crossentropy')

    # 训练模型
    model.fit([encoder_input_data,decoder_input_data],decoder_target_data,
              batch_size=batch_size,
              epochs=epochs,
              validation_split=0.2)
    model.save('s2s.h5')
    encoder_model.save('encoder_model.h5')
    decoder_model.save('decoder_model.h5')
```

模型训练结果如图12.11所示。

```
select train model or test model:train
训练模式..........
Train on 2400 samples, validate on 600 samples
Epoch 1/100
2400/2400 [==============================] - 27s 11ms/step - loss: 2.1924 - val_loss: 2.4341
Epoch 2/100
2400/2400 [==============================] - 24s 10ms/step - loss: 1.9873 - val_loss: 2.4850
Epoch 3/100
2400/2400 [==============================] - 23s 10ms/step - loss: 1.9516 - val_loss: 2.5290
Epoch 4/100
2400/2400 [==============================] - 25s 11ms/step - loss: 1.9198 - val_loss: 2.4833
Epoch 5/100
2400/2400 [==============================] - 19s 8ms/step - loss: 1.8801 - val_loss: 2.3805
Epoch 6/100
2400/2400 [==============================] - 18s 7ms/step - loss: 1.8343 - val_loss: 2.4091
Epoch 7/100
2400/2400 [==============================] - 35s 15ms/step - loss: 1.7853 - val_loss: 2.3898
Epoch 8/100
2400/2400 [==============================] - 28s 12ms/step - loss: 1.7338 - val_loss: 2.2818
Epoch 9/100
2400/2400 [==============================] - 18s 7ms/step - loss: 1.6941 - val_loss: 2.2460
Epoch 10/100
2400/2400 [==============================] - 26s 11ms/step - loss: 1.6549 - val_loss: 2.2657
```

图12.11 模型训练结果

```
Epoch 86/100
2400/2400 [==============================] - 17s 7ms/step - loss: 0.1858 - val_loss: 2.1673
Epoch 87/100
2400/2400 [==============================] - 15s 6ms/step - loss: 0.1810 - val_loss: 2.1552
Epoch 88/100
2400/2400 [==============================] - 15s 6ms/step - loss: 0.1766 - val_loss: 2.1769
Epoch 89/100
2400/2400 [==============================] - 18s 8ms/step - loss: 0.1721 - val_loss: 2.1618
Epoch 90/100
2400/2400 [==============================] - 14s 6ms/step - loss: 0.1665 - val_loss: 2.1986
Epoch 91/100
2400/2400 [==============================] - 15s 6ms/step - loss: 0.1634 - val_loss: 2.1776
Epoch 92/100
2400/2400 [==============================] - 21s 9ms/step - loss: 0.1597 - val_loss: 2.1814
Epoch 93/100
2400/2400 [==============================] - 15s 6ms/step - loss: 0.1546 - val_loss: 2.1995
Epoch 94/100
2400/2400 [==============================] - 14s 6ms/step - loss: 0.1547 - val_loss: 2.1976
Epoch 95/100
2400/2400 [==============================] - 14s 6ms/step - loss: 0.1469 - val_loss: 2.1961
Epoch 96/100
2400/2400 [==============================] - 14s 6ms/step - loss: 0.1441 - val_loss: 2.2048
Epoch 97/100
2400/2400 [==============================] - 21s 9ms/step - loss: 0.1430 - val_loss: 2.2013
Epoch 98/100
2400/2400 [==============================] - 15s 6ms/step - loss: 0.1370 - val_loss: 2.2206
Epoch 99/100
2400/2400 [==============================] - 14s 6ms/step - loss: 0.1360 - val_loss: 2.2281
Epoch 100/100
2400/2400 [==============================] - 14s 6ms/step - loss: 0.1304 - val_loss: 2.2186
```

图12.11 模型训练结果（续）

4. 模型测试

```
def test():
    encoder_model=load_model('encoder_model.h5', compile=False)
    decoder_model=load_model('decoder_model.h5', compile=False)
    ss=input("请输入要翻译的英文:")
    if ss=='-1':
        sys.exit()
    input_seq=np.zeros((1,INUPT_LENGTH,num_encoder_tokens))
    for t,char in enumerate(ss):
        input_seq[0,t,input_token_index[char]]=1.0
    decoded_sentence = decode_sequence(input_seq,encoder_model,decoder_model)
    print('-')
    print('Decoded sentence:', decoded_sentence)
```

5. 效果展示

```
if __name__ == '__main__':
    intro=input("select train model or test model:")
    if intro=="train":
        print("训练模式..........")
        train()
    else:
        print("测试模式.........")
        while(1):
            test()
```

模型测试结果如图 12.12 所示。

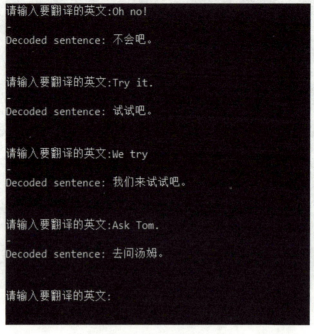

图12.12　模型测试结果

同学们可以调整参数，观察训练和测试效果。

单 元 小 结

　　传统的自然语言项目往往涉及多个自然语言处理模块的组合。这种作业方式存在严重的误差传播问题，亦即前一个模块产生的错误被输入到下一个模块中产生更大的错误，最终导致整个系统的脆弱性，为了解决这类问题，人们转向了另一种机器学习潮流的研究——深度学习。

　　本单元主要阐述了深度学习与自然语言处理中重要算法的发展、深度学习的主要算法流程以及主流算法模型，最终通过实战任务来实现机器翻译系统。本单元的实战属于综合性任务实战，会涉及前面单元的知识，通过本次实战，希望同学们能够灵活运用所学知识处理实际问题。

习　题

1．什么是深度学习？它在自然语言处理中有哪些应用？
2．深度学习模型中的循环神经网络是如何工作的？

参 考 文 献

[1] 宗成庆. 统计自然语言处理 [M]. 北京：清华大学出版社, 2013.
[2] 李航. 统计学习方法 [M]. 北京：清华大学出版社, 2012.
[3] 冯志伟. 中国术语标准化的由来与发展 [J]. 中国标准化, 2002(10):6-7.
[4] 荀恩东, 饶高琦, 肖晓悦, 等. 大数据背景下 BCC 语料库的研制 [J]. 语料库语言学, 2016,3(1): 93-109,118.
[5] 薛松. 汉英平行语料库中名词短语对齐算法的研究 [D]. 北京：中国科学院研究生院 (软件研究所), 2003.